PETER SINGER was born in Australia in 1946,
and educated at the University of Melbourne and
at Oxford, where he gained his first teaching
position. He subsequently taught at New York
University, has been a Woodrow Wilson Fellow,
and is now Professor and Chairman of the De-
partment of Philosophy at Monash University in
Australia. A contributor to *The New York Review
of Books, The New York Times Magazine, The
Nation,* and numerous philosophical journals, Mr.
Singer is also the author of *Democracy and Dis-
obedience, Animal Liberation, Practical Ethics,*
and *Marx.*

THE EXPANDING CIRCLE

Ethics and Sociobiology

BY
PETER SINGER

A MERIDIAN BOOK
NEW AMERICAN LIBRARY
TIMES MIRROR
NEW YORK AND SCARBOROUGH, ONTARIO

This publication was prepared under a grant from the
Woodrow Wilson International Center for Scholars, Washington, D.C.
The statements and views expressed herein are those of the author
and are not necessarily those of the Wilson Center

Designed by Kim Llewellyn

This is an authorized reprint of a hardcover edition published by
Farrar, Straus & Giroux, Inc.
The hardcover edition was published simultaneously in Canada by
McGraw-Hill Ryerson Ltd., Toronto.

SIGNET, SIGNET CLASSICS, MENTOR, PLUME, MERIDIAN
and NAL BOOKS are published *in the United States* by
The New American Library, Inc.,
1633 Broadway, New York, New York 10019,
in Canada by The New American Library of Canada Limited,
81 Mack Avenue, Scarborough, Ontario M1L 1M8

First Meridian Printing, April, 1982

1 2 3 4 5 6 7 8 9

PRINTED IN THE UNITED STATES OF AMERICA

For Renata

CONTENTS

PREFACE

Ethics is inescapable. Even if in grim adherence to some skeptical philosophy we deliberately avoid all moral language, we will find it impossible to prevent ourselves inwardly classifying actions as right or wrong. The skepticism that eschews all ethical judgment is possible only when all goes tolerably well: Nazi atrocities refute it more convincingly than volumes of philosophical argument.

Recognizing that we cannot do without standards of right and wrong is one thing; understanding the nature and origin of these standards is another. Is ethics objective? Are moral laws somehow part of the nature of the universe, like the laws of physics? Or are they of human origin? And if they are of human origin, are there standards of right and wrong that all human beings should accept, or must ethics always be relative to the society in which we live, perhaps even to the personal attitudes of each one of us?

Systematic Western philosophy goes back 2,500 years, and discussions of the nature of ethics date from the start of that period. Human beings have thought about these issues from the time they first began to inquire into the nature of their world and their society. Unlike our inquiries into the nature of the physical universe, however, two and a half millennia of moral philosophy have still not yielded generally accepted results about the fundamental nature of ethics.

For centuries, religion provided a way out of this difficulty. It is natural for those who believe in God to look to his wishes or commands for the origin of morality. By basing ethics on the will of God, believers did away with doubts about the objectivity and authority of morality.

One reason why religion no longer provides a satisfactory answer to the puzzle about the nature of morality is that religious belief itself is no longer as universally accepted as it once was. But there is also another problem in locating the origins of morality in the will of God. If all values result from God's will, what reason could God have for willing as he does? If killing is wrong *only* because God said: "Thou shalt not kill," God might just as easily have said: "Thou shalt kill." Would killing then have been right? To agree that it would have been right makes morality too arbitrary; but to deny that it would have been right is to assume that there are standards of right and wrong independent of God's will. Nor can the dilemma be avoided by claiming that God is good, and so could not have willed us to kill unjustly—for to say that God is good already implies a standard of goodness that is independent of God's decision. For this reason many religious thinkers now agree with the non-religious that the basis of ethics must be sought outside religion and independently of belief in God.

If religion cannot answer our worries about the nature of ethics, what of science? Ever since the experimental sciences began transforming what was once "natural philosophy" into what is now physics, there have been attempts to apply scientific methods to moral philosophy. The dream of a scientific ethics is an old one, but it has borne little fruit. Until quite recently it seemed to have died with Herbert Spencer and the Social Darwinists. Then in 1975 Edward O. Wilson, professor of zoology at Harvard University, published *Sociobiology: The New Synthesis*, a bold attempt to bring together biology, zoology, genetics, ethology, and studies of human behavior. In his opening paragraph Wilson claimed that the theory of natural selection should be pursued to explain ethics "at all depths." In the final chapter he suggested that the time may have come to take ethics away from the philosophers and hand it over to scientists.

I am a philosopher by training. Most of my colleagues in university departments of philosophy regard Wilson's invasion of their territory as too absurd to merit a considered response. It is true that the sociobiological approach to ethics often involves undeniable and crude errors. Nevertheless, I believe that the sociobiological approach to ethics does tell us something important about ethics, something we can use to gain a better understanding of ethics than has hitherto been possible. To show what this is, and how it can be combined with what is sound in philosophical theories of ethics, is the object of this book.

Most of this book was written while I was Fellow of the Woodrow Wilson International Center for Scholars, Washington, D.C. I thank the Center for its financial support and for the splendid facilities it provided for reading, thinking, and writing. I also thank Professor Ray Martin, Vice-Chancellor of Monash University, and Professor John Legge, Dean of the Faculty of Arts, for approving the unusually long period of leave which made it possible for me to accept the Wilson Center's offer of a fellowship. My wife, Renata, and my daughters, Ruth, Lee, and Esther, good-naturedly accepted the prolonged absence from home and friends that our stay in Washington involved.

More specific assistance came from a number of people: Edward O. Wilson took the time to read, annotate, and discuss with me an earlier draft. Other very helpful comments on the draft came from R. M. Hare and Richard Keshen. Penn Chu guided me into the literature on evolutionary biology and his diligent reading of the early chapters saved me from some errors in my account of current explanations of animal behavior. At several American universities I gave talks based on work in progress, and in return received comments too numerous to mention individually, but which collectively have had a major influence on the final product. The initial typing was efficiently coordinated by Eloise Doane, and done mostly by Miss Edith Ross; Jean Archer patiently retyped my numerous revisions after I returned to Monash.

Notes on the sources of quotations and of specific ideas mentioned in the text will be found at the back of the book.

Melbourne   Peter Singer
October 1980

The moral unity to be expected in different ages is not a unity of standard, or of acts, but a unity of tendency. . . . At one time the benevolent affections embrace merely the family, soon the circle expanding includes first a class, then a nation, then a coalition of nations, then all humanity, and finally, its influence is felt in the dealings of man with the animal world.

—W. E. H. LECKY, *The History of European Morals*

THE
EXPANDING
CIRCLE

1

THE ORIGINS OF ALTRUISM

We are not just rather like animals; we *are* animals.

—MARY MIDGLEY, *Beast and Man*

A NEW LOOK AT ETHICS

Human beings are social animals. We were social before we were human. The French philosopher Jean-Jacques Rousseau once wrote that in the state of nature human beings had "no fixed home, no need of one another; they met perhaps twice in their lives, without knowing each other and without speaking." Rousseau was wrong. Fossil finds show that five million years ago our ancestor, the half-human, half-ape creature known to anthropologists as *Australopithecus africanus,* lived in groups, as our nearest living relatives—the gorillas and chimpanzees—still do. As *Australopithecus* evolved into the first truly human being, *Homo habilis,* and then into our own species, *Homo sapiens,* we remained social beings.

In rejecting Rousseau's fantasy of isolation as the original or natural condition of human existence, we must also reject his account of the origin of ethics, and that of the school of social contract theorists to which he belonged. The social contract theory of ethics held that our rules of right and wrong sprang from some distant Foundation Day on which previously independent rational human beings came together to hammer out a basis for setting up the first human society. Two hundred years ago this seemed a plausible alternative to the then orthodox idea that morality represented the decrees of a divine lawgiver. It attracted some of the sharpest and most skeptical thinkers in Western social philosophy. If, however, we now know that we have lived in groups longer than we have been rational human beings, we can also be sure that we restrained our behavior toward our fellows before we were rational human beings. Social life requires some degree of restraint. A social grouping cannot stay together if its members make frequent and unrestrained attacks on one another. Just when a pattern of restraint toward other members of the group becomes a social ethic is hard to say; but ethics probably began in these pre-human patterns of behavior rather than in the deliberate choices of fully fledged, rational human beings.

Eighteenth-century philosophers like Rousseau had little information to draw on about the social behavior of non-human animals; and they knew even less about the evolution of human beings. Even after Darwin these topics were little studied, and what was known about animals came from the hostile perspective of the hunter, the exaggerated tales of adventurers, or accurate reports of the unnatural behavior of zoo animals. Only in recent years have both the study of animal behavior in the wild and the study of human evolution advanced to the point at which we can claim with some confidence to know something about ourselves and our animal ancestors and relatives. The most impressive attempt to bring all this new information together is Edward O. Wilson's huge book, *Sociobiology: The New Synthesis,* which appeared in 1975. Wilson defines sociobiology as "the systematic study of the biological basis of all social behavior." Since ethics is a form of social behavior—more than that, no doubt, but that at least—ethics falls within the scope of sociobiology. One might, of course, raise questions about the extent to which ethics has a biological basis; but if the origins of ethics lie in a past which we share with many non-human animals, evolutionary theory and observations of non-human social animals should have some bearing on the nature of ethics. So what does sociobiology offer us in place of the historical myth of the social contract?

Sociobiology bears on ethics indirectly, through what it says about the development of altruism, rather than by a direct study of ethics. Since it is difficult to decide when a chimpanzee or a gazelle is behaving ethically, this is a wise strategy. If we define altruistic behavior as behavior which benefits others at some cost to oneself, altruism in non-human animals is well documented. (This is not altruism in the usual sense, and in the next chapter we shall modify this definition; but for the moment it will do.) Understanding the development of altruism in animals will improve our understanding of the development of ethics in human beings, for our present ethical systems have their roots in the altruistic behavior of our early human and pre-human ancestors.

Altruism intrigues sociobiologists. Wilson calls it "the central theoretical problem of sociobiology." It is a problem because it has to be accounted for within the framework of Darwin's theory of evolution. If evolution is a struggle for survival, why hasn't it ruthlessly eliminated altruists, who seem to increase another's prospects of survival at the cost of their own?

ANIMAL ALTRUISM

Let us look at some examples of altruistic behavior in non-human animals. We can start with the warning calls given by blackbirds and thrushes when hawks fly overhead. These calls benefit other members of the flock, who can take evasive action; but giving a warning call presumably also gives away the location of the bird giving the call, thus exposing it to additional risk. (The calls are, acoustically, much harder to locate than other calls made by the same birds, but they must still make the bird easier to find than it would be if the bird were to hide itself without making any call at all.) If, as we would expect, birds who give warning calls are eaten at a higher rate than birds who act to save themselves without warning the rest of the flock, how does such altruism survive?

Another illustration comes from the behavior of Thomson's gazelles, a species of small antelope that is hunted by packs of African wild dogs. When a gazelle notices a dog pack, it bounds away in a curious, stiff-legged gait known as "stotting." Here is a description of this behavior and an indication of the puzzle it suggests:

Undoubtedly a warning signal it [stotting] spread wavelike in advance of the pack. Apparently in response to the stotting, practically every gazelle in sight fled the immediate vicinity. Adaptive as the warning display may seem, it nonetheless appears to have its drawbacks; for even after being singled out by the pack, every gazelle began the run for its life by stotting, and appeared to lose precious ground in the process . . . time and again we have watched the lead dog closing the gap until the quarry settled to its full running gait, when it was capable of making slightly better speed than its pursuer for the first half mile or so. It is therefore hard to see any advantage to the individual in stotting when chased, since individuals that made no display at all might be thought to have a better chance of surviving and reproducing.

Nor is altruism limited to warnings. Some animals threaten or attack predators to protect other members of their species. African wild dogs have been observed attacking a cheetah at considerable risk to their own lives in order to save a pup. Male baboons threaten predators, and cover the rear as the troop retreats. Parent birds frequently lead predators away from their nests with bizarre dances and displays which distract the predator's attention from the nest to the parent itself.

Food sharing is another form of altruism. Wolves and wild dogs bring meat back to members of the pack who were not in on the kill. Gibbons and chimpanzees without food gesture for, and usually receive, a portion of the food that another ape has. Chimpanzees also lead each other to trees with ripe fruit; indeed, their altruism extends beyond their own group, for when a whole group of chimpanzees is at a good tree, they make a loud booming noise which attracts other groups up to a kilometer away.

Several species help injured animals survive. Dolphins need to reach the surface of the water to breathe. If a dolphin is wounded so severely that it cannot swim to the surface by itself, other dolphins group themselves under it, pushing it upward to the air. If necessary they will keep doing this for several hours. The same kind of thing happens among elephants. A fallen elephant is likely to suffocate from its own weight, or it may overheat in the sun. Many elephant hunters have reported that when an elephant is felled, other members of the group try to raise it to its feet.

Finally, the restraint shown by many animals in combat

with their fellows might also be a form of altruism. Fights between members of the same social group rarely end in death or even injury. When one wolf gets the better of another, the beaten wolf makes a submissive gesture, exposing the soft underside of its neck to the fangs of the victor. Instead of taking the opportunity to rip out the jugular vein of his foe, the victor trots off, content with the symbolic victory. From a purely selfish point of view, this seems foolish. How is it that wolves who fight to kill, never giving a beaten enemy a second chance, have not eliminated those who pass up opportunities to rid themselves of their rivals forever?

EVOLUTION AND ALTRUISM

Many people think of evolution as a competition between different species; successful species survive and increase, unsuccessful ones become extinct. If evolution really worked mainly on the level of whole species, altruistic behavior between members of the same species would be easy to explain. The individual blackbird, taken by the hawk because of its warning call, dies to save the blackbird flock, thus increasing the survival prospects of the species as a whole. The wolf who accepts the submissive gesture of a defeated opponent exhibits an inhibition without which there would be no more wolves. And so on, for the other instances of altruism among animals.

The flaw in this simple explanation is that it is hard to see how, except under very special and rare conditions, the evolution of altruism could occur on so general a level as the survival or extinction of whole species. The real basis of selection is not the species, nor some smaller group, nor even the

individual. It is the gene. Genes are responsible for the characteristics we inherit. If a gene leads individuals to have some
feature which enhances their prospects of surviving and reproducing, that type of gene will itself survive into the next
generation; if a gene reduces the prospects of leaving offspring for those individuals who carry it, that type of gene
will itself die out with the death of the individual carrier.

For selection at the level of whole species to counteract
this individual selection of genes, evolution would have to select species at something like the rate at which it selects
genes. This means that old species would have to become extinct, and new species come into existence, nearly as often as
individuals either succeed or fail in reproducing. But of
course nature does not work like that; species evolve slowly,
over many, many generations. Hence any genes that lead to
altruism will normally lose out, in competition between
members of the *same* species, to genes that lead to more selfish behavior, before the altruistic genes could spread through
the species and so benefit the species as a whole in its competition with *other* species. And even if, under special circumstances, altruistic behavior did lead one species to survive where others without the genes for altruism became
extinct, competition within the species would still work
against the persistence of altruistic behavior in the surviving
species, once the external competition was over.

That, at least, is the broad account of evolution now accepted by many scientists working in this area. It is easy to
see how it undermines the simple account of the evolution of
altruism in terms of the survival of the species. Giving warning calls is a form of behavior with a genetic basis. Blackbirds
do not have to be taught to warn of predators. Now the question is: How could the genes for such self-sacrificing behavior

get established? How is it that, as soon as the combination of genes necessary for giving warning calls appears, this type of combination is not rapidly wiped out along with the individual birds who, by giving the warning, reduce their own prospects of living long enough to leave descendants? It may be true that if this happened the species as a whole would be less likely to survive; but all this shows is that there is a real puzzle as to how the species does survive, since the species as a whole is powerless to prevent the elimination of altruism within it.

The same problem arises in explaining other altruistic acts. Suppose that some wolves have genes which inhibit them against killing opponents who make submissive gestures, while other wolves, lacking these genes, finish off their defeated opponents. How will the inhibiting genes spread? If a killer wolf defeats an inhibited wolf in a fight, that will be the end of that particular set of inhibiting genes; if, on the other hand, an inhibited wolf defeats a killer wolf, the killer genes still survive and may reproduce. Over a long series of combats, it would seem that the killer genes ought to come to predominate among wolves. Why hasn't this happened?

Darwin himself was aware of this difficulty in the way of an evolutionary account of social and moral traits in humans. In *The Descent of Man* he wrote:

> But it may be asked, how within the limits of the same tribe did a large number of members first become endowed with these social and moral qualities, and how was the standard of excellence raised?
>
> It is extremely doubtful whether the offspring of the more sympathetic and benevolent parents, or of those who were the most faithful to their comrades, would be reared in greater numbers than the children of selfish and treacherous

parents belonging to the same tribe. He who was ready to sacrifice his life, as many a savage has been, rather than betray his comrades, would often leave no offspring to inherit his noble nature. The bravest men, who were always willing to come to the front in war, and who freely risked their lives for others, would on an average perish in larger numbers than other men. Therefore it hardly seems probable that the number of men gifted with such virtues, or the standard of their excellence, could be increased through natural selection, that is, by the survival of the fittest; for we are not here speaking of one tribe being victorious over another.

Darwin thought that part of the explanation was that as human reasoning powers increased, early humans would learn that if they helped their fellows, they would receive help in return; the remainder of his explanation was that virtuous behavior was fostered by the praise and blame of other members of the group. Sociobiologists do not invoke the institution of praise and blame for an explanation of altruism, since altruism occurs among non-human animals who do not praise or blame as we do. Sociobiologists have, however, developed Darwin's suggestion of the importance of the principle of reciprocity. They have suggested that two forms of altruism can be explained in terms of natural selection: kin altruism and reciprocal altruism. Some also allow a minor role for group altruism, but this is more controversial.

KIN ALTRUISM

Evolution can, as we have seen, be regarded as a competition for survival among genes. "Gene" as I use the term does not refer to the physical bits of DNA—which cannot survive any longer than the individual wolf, blackbird, or human in which they are present—but to the type of DNA. In this sense, genes can survive indefinitely, for one bit of DNA in one generation can lead to the existence of similar bits of DNA in the next. The most obvious way in which this can be done is by reproduction. Each sperm I produce contains a random sample of half my genes; therefore each time I fertilize an egg which grows into a child, a set of half my genes takes on an independent existence, with a chance of surviving my death and in turn passing some of its genes on down through the generations. So, for example, by "the gene for brown eyes" I do not mean the particular bit of biological matter I carry which will cause my child to have brown eyes; I mean the type of biological matter which, passed on in reproduction, leads human beings to have brown eyes.

Thus strictly selfish behavior—behavior aimed at furthering my own survival without regard for anyone else—will not be favored by evolution. I am doomed in any case. The survival of my genes depends largely on my having children, and on my children having children, and so forth. Evolution will favor, other things being equal, behavior which improves the prospects of my children surviving and reproducing.* Thus the first and most obvious way in which evolution can produce altruism is the concern of parents for their children. This is so widespread and natural a form of altruism that we do not usually think of it as altruism at all. Yet the

---

* I say "other things being equal" because under certain conditions there could be alternative strategies—like producing a larger number of children, and letting them take their chances. In mammals this option is not likely to be viable for females, since they must invest a lot of time in each offspring if any are to survive; but it could work for males, who can pass their genes on with much less labor. Sociobiologists argue that this accounts for the greater concern of females with child care, and the greater desire of males for casual sexual relationships with a variety of partners.

sacrifices that humans as well as many non-human animals constantly make for their children represent a tremendous effort for the benefit of beings other than themselves. Thus they must count as altruism, as we have defined the term so far. (In the case of humans, these sacrifices are well known to most parents, and to those who watch them; that they have not persuaded huge numbers of people against having children would be hard to explain if it were true that most people are selfish.)

So genes that lead parents to take care of their children are, other things being equal, more likely to survive than genes that lead parents to abandon their children. But taking care of one's children is only one way of increasing the chances of one's genes surviving. When I reproduce, my children do not have all the genes I have. (For that we will have to wait until we can clone genetic carbon copies.) Each child I produce contains half my genes; the other half of my children's genes comes, of course, from their mother. Each of my sisters and brothers will also, on average, have 50 percent of the same genes as I have, since, like me, they have half of my mother's and half of my father's genes. (This 50 percent is an average figure because, depending on how the genetic lottery fell out, they could have anything from all to none of their genes in common with me—but the huge number of genes involved makes either extreme fantastically unlikely.) Therefore in genetic terms my siblings are as closely related to me as my children; there is no special significance in the fact that the genes my children share with me replicate through my own body, whereas those I share with my sister did not. Assisting my brothers and sisters will enhance the prospects of my genes surviving, in much the same way as assisting my children will. (That care for siblings is not ordinarily as intense as care for offspring may be due to the fact that the difference in age makes parents able to care for their offspring when the offspring most need it, whereas siblings usually are too young to do so. In addition, in non-monogamous species full siblings are the exception, and half siblings—where the genetic relationship is only 25 percent—the rule.)

This is the basis of kin altruism: the genetically based tendency to help one's relatives. The relationship does not have to be as close as that of parents to their children or siblings to each other. The proportion of genes in common does fall off sharply as it becomes more distant—between aunts (or uncles) and their nieces (or nephews) it is 25 percent; between first cousins 12½ percent—but what is lacking in quality can be made up for by an increased quantity. Risking my life will not harm the prospects of my genes surviving if it eliminates a similar risk to the lives of two of my children, four of my nieces, or eight of my first cousins. Thus kin selection can explain why altruism should extend beyond the immediate family. In close-knit groups, where most members are related to other members, kin selection may explain altruistic behavior like giving the alarm when predators are near, which benefits the entire group.

Kin altruism does not imply that animals know how closely related they are to each other—that they can distinguish full sisters from half sisters, or cousins from unrelated animals. The theory says only that animals can be expected to act roughly *as if* they were aware of these relationships. In fact, since we are talking about complex living beings, there are many instances where animals do not behave in accordance with the nicely calculated fractions of genetic relationships. A female chimpanzee with many reproductive years ahead of her may sacrifice her life for a single child. African wild dogs have been observed risking their lives by attacking a cheetah that was threatening a pup that was at most a nephew to them. Evolved behavioral tendencies are not as predictable as the motions of the planets. Nevertheless, kin selection can explain some otherwise mysterious facts. For instance, why do adult zebras defend any calf in the herd attacked by a predator, whereas wildebeest do not? The reason could be that zebras live in family groups, so that adults and calves would generally all be related; wildebeest interbreed much more with other groups and adults would not be related to randomly selected calves. More startling still is the infanticide practiced by male langur monkeys. Female langurs live in groups, each under the control of a dominant male who prevents any other male from breeding with them. The other males, being unwilling bachelors, try to overthrow the dominant male and take his harem. If one should succeed, he will set about killing all the infants in his newly acquired group. This may not be good for the species as a whole, but the killer is not related to his victims; moreover, females nursing infants do not ovulate, so by removing the infants the male is able to have his own children earlier than would otherwise be possible. To these children he will be a better father. In the difference between his behavior toward infants genetically related to him and his behavior toward those that are not, the langur monkey demonstrates in a brutally clear form the kind of "altruism" that may evolve through kin selection. (Male lions have also been observed to kill infants on taking over a pride. Is there a human parallel in the wicked stepparents so common in fairy tales? Or in the mass rapes that for centuries have characterized military conquests?)

RECIPROCAL ALTRUISM

Kin altruism exists because it promotes the survival of one's relatives; but not all altruistic acts help relatives. Monkeys spend a lot of their time grooming each other, removing parasites from those awkward places a monkey cannot itself reach. Monkeys grooming each other are not always related. Here reciprocal altruism offers an explanation: you scratch my back and I'll scratch yours.

Here's another example: I see a stranger drowning and I jump in to save her. Suppose that in so doing I run a 5 percent risk of drowning myself; suppose too that without my help the stranger would run a 50 percent risk of drowning, but that with my help she will be saved, except in the 5 percent of cases in which we both drown. At first glance, jumping in seems to be a purely altruistic act. I run a 5 percent risk of death in order to help a stranger. But suppose that one day I myself will need to be rescued, and the person I saved this time will then jump in and help me. Suppose that without help I would have a 50 percent chance of drowning, but with help my prospects improve to 95 percent. Then, taking the two acts together, it is in my interest to save the drowning stranger, for I thus exchange two separate small risks (the 5 percent risk I incur when I help the stranger and also when I am helped) for one large risk (the 50 percent risk I would have if I were not helped). Obviously two 5 percent risks are better than one 50 percent risk.

This is an artificial example, with the risks made precisely measurable in order to make the benefit clear. One might question the example on that ground; but there is a more important question that needs to be asked about the example: What is the link between rescuing a stranger and being rescued oneself? If one can arrange to get rescued without having to do any rescuing oneself, that seems the best strategy, from a self-interested standpoint. Why isn't that what happens? What ensures that this form of altruism is reciprocal?

On one level, the answer to this question could be that individuals can remember who has helped them and who has not, and they will not help anyone who has refused to help them. Cheats—those who take help but refuse to give it— never prosper, for their cheating is noticed and punished. If this is right we would expect reciprocal altruism only among creatures capable of recognizing other individuals, sorting them into those who help and those who do not. Reciprocity may not require human reasoning powers, but it would require intelligence. It would also be more likely in species with a relatively long life span, living in small, stable groups. For in this way, opportunities for repeated reciprocal acts would occur more frequently.

The evidence supports this conclusion. Reciprocal altruism is most common among, and perhaps limited to, birds and mammals; its clearest cases come from highly intelligent social animals like wolves, wild dogs, dolphins, baboons, chimpanzees, and human beings. In addition to grooming each other, members of these species often share food on a reciprocal basis and help each other when threatened by predators or other enemies.

On another level, there is still a problem: How did this reciprocal altruism get going? After all, reciprocal altruism looks rather like the social contract model of ethics, which we have already dismissed as a historical fantasy—and the idea of a contract becomes even more fantastic if it is extended to non-human animals. But if there was no deliberate contract of the "you scratch my back and I'll scratch yours" kind, the first animals to risk their lives for other, unrelated members of their species were risking their lives without much prospect of anything in return. If reciprocal altruism is widely practiced, it pays to take part—chances are, you'll benefit later. But if reciprocal altruism is rare, it might be better, from the self-interested point of view, not to put yourself out. In the drowning example just given, it would not pay to rescue another, running a 5 percent risk of drowning oneself, unless by doing so one significantly raises the chances that one will oneself be rescued when the need arises. So it is not quite true that cheats never prosper. Cheats prosper until there are enough who bear grudges against them to make sure they do not prosper. If we imagine a group consisting partly of those who accept help but give none—"cheats"—and partly of those who accept help and give help to all except those who have refused to help—call them "grudgers"—there is a critical number of how many grudgers there must be before it pays to be a grudger rather than a cheat. One grudger in a population of cheats will get cheated often and never be helped; but the more grudgers and fewer cheats there are, the more often the grudger will be repaid for her help and the more rarely she will be cheated. So while we can understand why reciprocal altruism should prosper after it gets established, it is less easy to see why the genes leading to this form of behavior did not get eliminated as soon as they appeared.

GROUP ALTRUISM

It may be that to explain how reciprocal altruism can get established, we need to allow a limited role for a form of group selection. Imagine that a species is divided into several iso-lated groups—perhaps they are monkeys whose terrain is di-vided by rivers which, except in rare droughts, are too swift to cross. Now suppose that reciprocal altruism somehow ap-pears from time to time in each of these groups. Let us say that one monkey grooms another monkey, searching for dis-ease-carrying parasites; when it has finished it presents its own back to be groomed. If the genes that make this behavior probable are rare mutations, in most cases the altruistic mon-key would find its kindness unrewarded; the groomed mon-key would simply move away. Grooming strangers would therefore bring no advantage, and since it leads the monkey to spend its time helping strangers instead of looking after it-self, in time this behavior would be eliminated. This elimina-tion may not be good for the group as a whole, but as we have seen, within the group it is individual rather than group se-lection that dominates.

Now suppose that in one of these isolated groups it just happens that a lot of monkeys have genes leading them to in-itiate grooming exchanges. (In a small, closely related group, kin altruism might bring this about.) Then, as we have seen, those who reciprocate could be better off than those who do not. They will groom and be groomed, remaining healthy while other members of the group succumb to the parasites. Thus in this particular isolated group, possessing the genes for reciprocal grooming will be a distinct advantage. In time, all the group would have them.

There is one final step. The reciprocal grooming group now has an advantage, as a group, over other groups who do not have any way of ridding themselves of parasites. If the parasites get really bad, the other groups may become ex-tinct, and one dry summer the pressure of population growth in the recripocal grooming group will push some of its members across the rivers into the territories formerly occupied by the other groups. In this way group selection could have a limited role—limited because the required conditions would not often occur—in the spread of reciprocal altruism.

If we are prepared to allow group selection a role in the inception of reciprocal altruism, we can hardly deny that the survival of some groups rather than others can provide an evolutionary explanation for a more general tendency for altruistic behavior toward other members of a group. This is still quite distinct from the popular view of traits evolving because they help the species survive—groups are far smaller units than species, and come in and out of existence much more frequently, so group selection is more likely to be an effective counterweight to individual selection than is species selection. Nevertheless, a group would have to keep itself distinct from other groups for group altruism to work—otherwise more egoistically inclined outsiders would work their way into the group, taking advantage of the altruism of members of the group without offering anything in return. They would then outbreed the more altruistic members of the group and so begin to outnumber them, until the group would cease to be more altruistic than any other group of the same species. Although this would cost it its evolutionary advantage over other groups, there would be no mechanism for stopping this. If the group altruism had been essential to the group's survival, the group would simply die out.

This suggests that group altruism would work best when coupled with a degree of hostility to outsiders, which would protect the altruism within the group from penetration and subversion from outside. Hostility to outsiders is, in fact, a very common phenomenon in social animals. Although there is a popular myth that human beings are the only animals who kill members of their own species, other species can be as unpleasant toward foreigners as we are. Many social animals, from ants through chickens to rats, will attack and often kill outsiders placed in their midst. In a series of experiments conducted on rhesus monkeys, it has been shown that introducing a strange rhesus monkey into an established group aroused much more aggression than either crowding the monkeys or reducing their food supply. Admittedly, keeping strangers away could just be a means of protecting one's own food supply and that of one's kin; but it could also be that this behavior serves the same role as geographical isolation in protecting the altruism of the group from debasement.

It may be objected that in a small, isolated group of the kind I have described, there will be so much interbreeding that all members of the group will be related to each other, and so what we have is not group selection at all, but rather kin selection in the special case in which all the group are kin to each other. This may be so; certainly kin and non-kin selection will be hard to distinguish in this situation. Nevertheless, when members of the group behave in certain ways toward all other members of the group—irrespective of whether they are full siblings or very distant cousins—and when this behavior gives the entire group a selective advantage over other groups, it is reasonable to describe what is going on as "group selection" even if it may ultimately be possible to explain what is going on in terms of kin selection.

Keeping outsiders away would not be enough to prevent erosion of high levels of self-sacrificing behavior for the benefit of the group. Evolutionary theory would lead us to expect a drift back toward selfishness within the group, since individuals who behaved selfishly would reap the benefits of the sacrifices of others without making any sacrifices themselves. Perhaps, though, a group could develop a way of dealing with a small number of free-riders who emerge within it. Human societies, at least, have institutions which serve this end; but here we are beginning to look beyond the development of altruism in non-human animals to its existence in our own species.

2

THE BIOLOGICAL
BASIS OF ETHICS

> We should all agree that each of us is bound to show
> kindness to his parents and spouse and children, and to
> other kinsmen in a less degree; and to those who have
> rendered services to him, and any others whom he may
> have admitted to his intimacy and called friends; and to
> neighbours and to fellow-countrymen more than others;
> and perhaps we may say to those of our own race more
> than to black or yellow men, and generally to human
> beings in proportion to their affinity to ourselves.
>
> —HENRY SIDGWICK, *The Methods of Ethics*

Every human society has some code of behavior for its members. This is true of nomads and city-dwellers, of hunter-gath-
erers and of industrial civilizations, of Eskimos in Greenland
and Bushmen in Africa, of a tribe of twenty Australian aborigines and of the billion people that make up China. Ethics
is part of the natural human condition.

That ethics is natural to human beings has been denied.
More than three hundred years ago Thomas Hobbes wrote in
his *Leviathan:*

During the time men live without a common Power to keep
them all in awe they are in that condition called War; and

such a war, as is of every man against every other man. . . . To this war of every man against every man, this also is consequent; that nothing can be Unjust. The notions of Right and Wrong, Justice and Injustice have there no place.

Hobbes's guess about human life in the state of nature was no better than Rousseau's idea that we were naturally solitary. It is not the force of the state that persuades us to act ethically. The state, or some other form of social power, may reinforce our tendency to observe an ethical code, but that tendency exists before the social power is established. The primary role Hobbes gave to the state was always suspect on philosophical grounds, for it invites the question why, having agreed to set up a power to enforce the law, human beings would trust each other long enough to make the agreement work. Now we also have biological grounds for rejecting Hobbes's theory.

Occasionally there are claims that a group of human beings totally lacking any ethical code has been discovered. The Ik, a northern Uganda tribe described by Colin Turnbull in *The Mountain People,* is the most recent example. The biologist Garrett Hardin has even claimed that the Ik are an incarnation of Hobbes's natural man, living in a state of war of every Ik against every other Ik. The Ik certainly were, at the time of Turnbull's visit, a most unfortunate people. Originally nomadic hunters and gatherers, their hunting ground was turned into a national park. They were forced to become farmers in an arid mountain area in which they had difficulty supporting themselves; a prolonged drought and consequent famine was the final blow. As a result, according to Turnbull, Ik society collapsed. Parents turned their three-year-old children out to fend for themselves, the strong took food from the mouths of the weak, the sufferings of the old and sick were a source of laughter, and anyone who helped another was considered a fool. The Ik, Turnbull says, abandoned family, cooperation, social life, love, religion, and everything else except the pursuit of self-interest. They teach us that our much vaunted human values are, in Turnbull's words, "luxuries that can be dispensed with."

The idea of a people without human values holds a certain repugnant fascination. *The Mountain People* achieved a rare degree of fame for a work of anthropology. It was reviewed in *Life*, talked about over cocktails, and turned into a stage play by the noted director Peter Brook. It was also severely criticized by some anthropologists. They pointed out the subjective nature of many of Turnbull's observations, the vagueness of his data, contradictions between *The Mountain People* and an earlier report Turnbull had published (in which he described the Ik as fun-loving, helpful, and "great family people"), and contradictions within *The Mountain People* itself. In reply Turnbull admitted that "the data in the book are inadequate for anything approaching proof" and recognized the existence of evidence pointing toward a different picture of Ik life.

Even if we take the picture of Ik life in *The Mountain People* at face value, there is still ample evidence that Ik society has an ethical code. Turnbull refers to disputes over the theft of berries which reveal that, although stealing takes place, the Ik retain notions of private property and the wrongness of theft. Turnbull mentions the Ik's attachment to the mountains and the reverence with which they speak of Mount Morungole, which seems to be a sacred place for them. He observes that the Ik like to sit together in groups and insist on living together in villages. He describes a code that has to be followed by an Ik husband who intends to beat his wife, a code that gives the wife a chance to leave first. He reports that the obligations of a pact of mutual assistance known as *nyot* are invariably carried out. He tells us that there is a strict prohibition on Ik killing each other or even drawing blood. The Ik may let each other starve, but they apparently do not think of other Ik as they think of any non-human animals they find—that is, as potential food. A normal well-fed reader will take the prohibition of cannibalism for granted, but under the circumstances in which the Ik were living human flesh would have been a great boost to the diets of stronger Ik; that they refrain from this source of food is an example of the continuing strength of their ethical code despite the crumbling of almost everything that had made their lives worth living.

Under extreme conditions like those of the Ik during famine, the individual's need to survive becomes so dominant that it may seem as if all other values have ceased to matter, when in fact they continue to exercise an influence. If any conditions can be worse than those the Ik endured, they were the conditions of the inmates of Soviet labor camps and, more horrible still, the Nazi death camps. Here too, it has been said that "the doomed devoured each other," that "all trace of human solidarity vanished," that all values were erased and every man fought for himself. Nor should it be surprising if this were so, for the camps deliberately and systematically dehumanized their inmates, stripping them naked, shaving their hair, assigning them numbers, forcing them to soil their clothing with excrement, letting them know in a hundred ways that their lives were of no account, beating them, torturing them, and starving them. The astonishing thing is that despite all this, life in the camps was *not* every man for himself. Again and again, survivors' reports show that prisoners helped each other. In Auschwitz prisoners risked their lives to pick up strangers who had fallen in the snow at roll call; they built a radio and disseminated news to keep up morale; though they were starving, they shared food with those still more needy. There were also ethical rules in the camps. Though theft occurred, stealing from one's fellow prisoners was strongly condemned and those caught stealing were punished by the prisoners themselves. As Terrence Des Pres observes in *The Survivor,* a book based on reports by those who survived the camps: "The assumption that there was no moral or social order in the camps is wrong. . . . Through innumerable small acts of humanness, most of them covert but everywhere in evidence, survivors were able to maintain societal structures workable enough to keep themselves alive and morally sane."

The core of ethics runs deep in our species and is common to human beings everywhere. It survives the most appalling hardships and the most ruthless attempts to deprive human beings of their humanity. Nevertheless, some people resist the idea that this core has a biological basis which we have inherited from our pre-human ancestors. One ground for resistance is that we like to think of our own actions as radically different from the behavior of animals, no matter how altruistic those animals may be. Animals act instinctively; humans are rational, self-conscious beings. We can reflect on the rightness or wrongness of our actions. Animals cannot. We can follow moral rules. We can see what is good, and choose it. Animals cannot. Or so many people think.

Attempts to draw sharp lines between ourselves and other animals have always failed. We thought we were the only beings capable of language, until we discovered that chimpanzees and gorillas can learn more than a hundred words in sign language, and use them in combinations of their own devising. Scientists are now laboriously discovering what many dog owners have long accepted; we are not the only animals that reason. As Darwin wrote in *The Descent of Man:* "The difference in mind between man and the higher animals, great as it is, certainly is one of degree and not of kind." It is a mistake to think of all animals as doing by blind instinct what we do by conscious deliberation. Both human and non-human animals have innate tendencies toward behaving in particular ways. Some of these tendencies rigidly prescribe a particular kind of behavior—like the fly, so set on going in one direction that it buzzes repeatedly into the glass, instead of trying different directions until it comes to the part of the window that is open. Other innate tendencies merely set a goal which leaves room for a diversity of strategies—like the fox that "instinctively" wants a hen and, as those who keep hens learn to their cost, can think of dozens of different ways to get it. The "instincts" of the social mammals are mostly of this more open sort. In this sense human beings have "instincts" too: think about how hard it is for parents to hear their baby cry without picking it up, or for adolescent and older humans to avoid taking an interest in sex.

Another ground for resisting the idea that ethics has a biological basis is that ethics is widely regarded as a cultural phenomenon, taking radically different forms in different societies. As our knowledge of remoter parts of the globe increased, so too did our awareness of the variety of human ethical codes. Edward Westermarck's *The Origin and Development of the Moral Ideas*, published in 1906–8, consists of two large volumes, a total of more than 1,500 pages, comparing differences among societies on such matters as the wrongness of killing (including killing in warfare, euthanasia, suicide, infanticide, abortion, human sacrifices, and duels); whose duty it is to support children, or the aged, or the poor; the position of women, and the forms of sexual relations permitted; the holding of slaves, the right to property in general, and what constitutes theft; the duty to tell the truth; dietary restrictions; concern for non-human animals; duties to the dead, and to the gods; and so on. The overwhelming impression we get from Westermarck's book, and from most anthropological literature, is of an immense diversity in ethics, a diversity which must be of cultural rather than biological origin. Edward O. Wilson has conceded: "The evidence is strong that almost all differences between human societies are based on learning and social conditioning, rather than heredity." So it may seem that if we want to discuss human ethics we must shift our attention from biological theories of human nature to particular cultures and the factors that have led them to develop their own particular ethical codes. Yet while the diversity of ethics is indisputable, there are common elements underlying this diversity. Moreover, some of these common elements are so closely parallel to the forms of altruism observable in other social animals that they render implausible attempts to deny that human ethics has its origin in evolved patterns of behavior among social animals. I shall start with the ethical form of kin altruism.

KIN SELECTION IN HUMAN ETHICS

The Methods of Ethics, from which I have taken the quotation at the beginning of this chapter, is a philosophical treatise on ethics written by the Cambridge philosopher Henry Sidgwick and first published in 1874. The passage quoted is a description of the principles regulating the duty of benevolence, as this duty was generally understood at that time, rather than a statement of Sidgwick's own views. It gives a graduated list of those to whom we should be kind which fits neatly with sociobiological theories. First place goes to kin altruism; then come reciprocal and group altruism. In this respect late Victorian England was not unusual. As Westermarck notes in *The Origin and Development of the Moral Ideas*, a mother's duty to look after her children has seemed so obvious that most anthropological accounts scarcely bother to mention it. The duty of a married man to support and protect his family is, Westermarck says, equally widespread, and he backs up his claim with a score of examples. His account of the duties almost universally accepted among human beings parallels Sidgwick's list in placing duties to parents alongside duties to children and wives, with duties to aid brothers and sisters closely following, and those toward more distant relatives more variable, but still prominent in most societies. Benevolence to other members of the tribe or group comes next in importance, with benevolence to outsiders often lacking entirely.

The universal importance of kinship in human societies has been recognized by one of sociobiology's strongest critics, the anthropologist Marshall Sahlins. In *The Use and Abuse of Biology*, Sahlins has written: "Kinship is the dominant structure of many of the peoples anthropologists have studied, the prevailing code not only in the domestic sphere but generally of economic, political and ritual action." Sahlins goes on to deny that this has a biological basis, pointing out that who is recognized as "kin" in different cultures often fails to correspond to strict degrees of blood relationship. Sahlins, however, takes the sociobiological thesis too narrowly. His examples show that there is generally a considerable correlation between blood relationship and acceptance of someone as "kin." There is no need for a sociobiologist to demand more. Any reasonable sociobiologist will admit that culture plays a role in the structure of human societies, and biological forces therefore cannot always take the shortest and simplest route.

Today the unconcealed racial distinction referred to in Sidgwick's final sentence strikes a discordant note; but the reality of Sidgwick's account of the degrees of benevolence holds remarkably well, considering how many other elements of Victorian morality have changed in the past century. We still think first of our immediate family, then of friends, neighbors, and more distant relatives, next of our fellow citizens generally, and last of all of those who have nothing in common with us except that they are human beings. Think of our reaction to news of a famine in Africa. Those of us who care at all may send a donation to one of the agencies trying to help: ten dollars, or fifty dollars, or perhaps even a hundred dollars. Any more would be a rare act of generosity by the standards of our society. Yet those of us fortunate enough to live in Western Europe, North America, Australia, or Japan regularly spend as much or more on holidays, new clothes, or presents for our children. If we cared about the lives and welfare of strangers in Africa as we do about our own welfare and that of our children, would we spend money on these nonessential items for ourselves instead of using it to save lives? Of course, we have lots of excuses for not sending money to Africa: we say that our contribution could only be a drop in the ocean, or that the agencies waste the money they receive, or that food handouts are no good—what is needed is development, or a social revolution, or population control. In our more honest moments, though, we recognize that these are excuses. My contribution cannot end a famine, but it can save the lives of several people who might otherwise starve. While we seize on every newspaper report of relief efforts being wasted as a justification for not giving, how many of us bother to look at the overall efficiency of aid organizations, which is, in the case of voluntary organizations, actually very high by the standards of large corporations? And if we think that not food aid, but development, or revolution, or population control is the real answer to the famine problem, why aren't we contributing to groups promoting these solutions?

I have written and lectured on the subject of overseas aid, arguing that our affluence puts us under an obligation to do much more than we are now doing to help people in real need. At the popular level, the most frequent response is that we should look after our own poor first. Among philosophers, essentially similar replies are put in a more sophisticated manner. For instance, I have been told that while we should certainly do more to prevent poverty, we ought to do nothing inconsistent with our obligation to do the best for our children—an obligation which, it turns out, includes sending them to expensive private schools and buying them ten-speed bicycles. The proposal that we might risk lessening the happiness or prospects of our own children, to however slight a degree, in order to save strangers from starvation strikes many people as not merely idealistic but positively wrong.

The preference for "our own" is understandable in terms of our evolutionary history. It is an instance of the kin altruism we observed in other animals, with an element of group altruism added. This does not mean that a society must encourage its members to act in accordance with this preference. There have recently been concerted attempts to eliminate certain forms of preference for members of one's own group. In a multiracial society, preferences for members of one's own race or ethnic group often lead to strife, and in many countries it is now considered wrong to prefer those of one's own race or ethnic group in employment, education, or housing. Sanctions are invoked against those who do it. These efforts toward racial equality meet strong resistance, as one would expect from any attempt to counter a deep-seated bias, but they have generally been successful in changing people's attitudes, as well as their actions, toward fellow citizens of different races and ethnic backgrounds.

It has long been the dream of social reformers to carry out a similar equalizing process in respect of the family, so that members of a community no longer automatically prefer the interests of members of their family to those of the community. Like so many other perennial ideas, this goes back to Plato. In the *Republic,* Plato argues that unity is the greatest good in a community, and unity occurs "where all the citizens are glad or grieved on the same occasions of joy and sorrow." To bring this about, at least among the Guardians who are the rulers of Plato's ideal state, he suggests that there should be no separate households or marriages, but a form of communal marriage. Thus instead of "each man dragging any acquisition which he has made into a separate house of his own, where he has a separate wife and children and private pleasures and pains," we will have a situation in which the Guardians "are all of one opinion about what is near and dear to them, and therefore they all tend towards a common end."

In his estimate of the divisive effect of the family within a strongly collective community, Plato was right. Yonina Talmon unconsciously echoes Plato in her sociological study of Jewish collective settlements, *Family and Community in the Kibbutz.* Referring to the early pioneering days of these set-

tlements, when the need for unity was strongest, Talmon
says:

> Family ties are based on an exclusive and discriminating loy-
> alty which sets the members of one's family more or less apart
> from others. Families may easily become competing foci of
> emotional involvement that can infringe on loyalty to the
> collective. Deep attachment to one's spouse and children . . .
> may gain precedence over the more ideological and more
> task-oriented relations with comrades.

But Plato's optimism over the prospects of doing away with
the family has not been borne out by subsequent experiments. For the reasons Talmon gives, the kibbutz movement
began by strongly discouraging family attachments. Children
lived together in communal houses, apart from their parents.
From infancy the kibbutz provided nurses and teachers, free-
ing both parents for work. Meals were taken in a communal
dining room, not in family units, and entertainment was communal. Children were encouraged to call their parents by
their names, rather than "father" or "mother." There was a
ban on couples working in the same place, and husbands and
wives who spent most of their time together were viewed
with scorn. Upheld by deep ideological commitment to socialism and to Jewish settlement in what was then Palestine,
these extreme limits on family relations were accepted as
part of the settlement's struggle against an external environment that was hostile both in respect of the difficulties of
growing food and in respect of the surrounding Palestinian
population. In humans as well as other animals, an external
threat leads a group to be more cohesive than usual. (Compare the sacrifices readily made by all sections of the British
population during the Second World War, when there was a
real external threat, with the lack of response to appeals by

successive British Prime Ministers for a return to "the spirit of Dunkirk" to halt Britain's economic decline.) When the survival of the entire group is at stake, our concern with our own interests is subordinated to the need to ensure that the group survives; but in more normal times, individual interests return. Perhaps groups not able thus to mobilize resources in emergencies did not survive; while individuals who did not press their own interests when the emergency was over passed on fewer descendants.

In any case, for whatever reason, once the kibbutzim became established, and the twin dangers of starvation and Arab attack receded, the ideal of unity proved insufficient to maintain the original intensity of communal feeling. The kibbutzim have survived, but they have had to come to terms with the family. Couples spend more time in their apartments, often taking their meals there; children are with their parents for much of their free time, and often sleep in the family apartment rather than the children's house; it is no longer frowned upon for couples to sit next to each other on all public occasions; children once again call their parents "father" and "mother."

The experience of the kibbutz movement parallels that of other attempts to make the community, instead of the family, the basic unit of concern. After the Bolshevik Revolution, attempts were made in the Soviet Union to carry out the call of the *Communist Manifesto* for the abolition of the family. Within twenty years Soviet policy swung around completely, and began to encourage family life. (The *Manifesto* itself, incidentally, is equivocal about abolishing the family; not surprisingly, since Marx was as devoted to his family as any father.) Some religious communities have begun by bringing up children collectively; but the family has bounced back as soon as the bloom of spiritual enthusiasm fades. Monastic settlements have achieved a more permanent suppression of the family, but a community based on strict celibacy is hardly viable on its own.

A bias toward the interests of our own family, rather than those of the community in general, is a persistent tendency in human behavior, for good biological reasons. Not every persistent human tendency, however, is universally regarded as a virtue. (Compare attitudes toward another persistent human tendency, which probably also has a biological basis, the tendency to have sexual relations with more than one partner.) Why is it that in almost every human society concern for one's family is a mark of moral excellence? Why do societies not merely tolerate but go out of their way to praise parents who put the interests of their children ahead of the interests of other members of the community? The answer may lie, not just in the universality and strength of family feeling, but also in the benefits to society as a whole that come from families taking care of themselves. When families see that the children are fed, kept clean and sheltered, that the sick are nursed and the elderly cared for, they are led by bonds of natural affection to do what would otherwise fall on the community itself and either would not be done at all or would require labor unmotivated by natural impulses. (In a large modern community, it would require an expensive and impersonal bureaucracy.) Given the much greater intensity of family feeling compared with the degree of concern we have for the welfare of strangers, ethical rules which accept a degree of partiality toward the interests of one's own family may be the best means of promoting the welfare of all families and thus of the entire community.

RECIPROCAL ALTRUISM
AND HUMAN ETHICS

Though kinship is the most basic and widespread bond between human beings, the bond of reciprocity is almost as universal. In his description of the Victorian moral view, Sidgwick lists a person's duty to show kindness "to those who have rendered services to him" immediately after the duty to be kind to kinsmen. Among the Ik, the mutual-assistance pact known as *nyot* survived when the family itself was breaking up. Westermarck says: "To requite a benefit, or to be grateful to him who bestows it, is probably everywhere, at least under certain circumstances, regarded as a duty." Since Westermarck wrote, anthropologists from Marcel Mauss to Claude Lévi-Strauss have continued to stress the importance of reciprocity in human life. Howard Becker, author of *Man in Reciprocity*, finds our tendency for reciprocity so universal that he has proposed renaming our species *Homo reciprocus*. After surveying these and other recent studies, the sociologist Alvin Gouldner has concluded: "Contrary to some cultural relativists, it can be hypothesized that a norm of reciprocity is universal."

It is surprising how many features of human ethics could have grown out of simple reciprocal practices like the mutual removal of parasites from awkward places that one cannot oneself reach. Suppose I want to have the lice in my hair picked out. To obtain this I am willing to pick out someone else's lice. I must, however, be discriminating in selecting whom to groom. If I help everyone indiscriminately I shall find myself grooming others who do not groom me back. To avoid this waste of time and effort I distinguish between those who repay me for my assistance and those who do not.

In other words, I separate those who deal fairly with me from those who cheat. Those who do not repay me I shall mark out to avoid; indeed, I may go further still, reacting with anger and hostility. Conceivably it will benefit me and other reciprocating altruists in my group if we make sure that the worst "cheats" are unable to take advantage of any of us again; killing them or driving them away would be effective ways of doing this. For those who do all that I hope they will do, on the other hand, I will have a positive feeling that increases the likelihood of my doing my part to preserve and develop a mutually advantageous relationship.

Let us take individually these outgrowths of reciprocal altruism. The first and most crucial is the distinction between those it is worth my while to assist and those it is not. Of course, if we all wait for each other to begin, we shall never get going. Initially someone has to remove someone else's parasites without knowing if there will be a return. After a bit of this, however, the track record of each member of the group will become clear. Then I can stop helping those who have not helped me. This requires a sense of what amounts to sufficient repayment for the help I have given. If I take an hour meticulously removing every louse from someone else's head, and she refuses even to look at my head, the verdict is clear; but what if she hurries over my head in ten minutes, leaving at least some of my lice in place? No doubt the practice of reciprocal altruism can tolerate rough justice at this point, but we would expect that as human powers of reasoning and communicating increased, decisions as to what is or is not an equitable exchange would become more precise. They would begin to take into account variations in circumstances: If, for instance, I can remove your few lice in ten minutes, should I demand that you spend the hour it would take to get rid of the multitude on my scalp? In answering this kind of question we would begin to develop a concept of fairness. More than two thousand years ago the Greek historian Polybius observed:

> ... when a man who has been helped when in danger by another does not show gratitude to his preserver, but even goes to the length of attempting to do him injury, it is clear that those who become aware of it will naturally be displeased and offended by such conduct, sharing the resentment of their injured neighbor and imagining themselves in the same situation. From all this there arises in everyone a notion of the meaning and theory of duty, which is the beginning and end of justice.

To say that the duty to repay benefits is the beginning and end of justice is an overstatement; but that it is the beginning is plausible. To "repay benefits" we should add the converse, "revenge injuries"; for the two are closely parallel and generally seen as going together. In tribal ethics the duties of gratitude and revenge often have a prominence they lack in our culture today. (I am not saying that they are not important motives in our society. They are; but we are less likely now to praise vengefulness as a virtue, and even gratitude no longer ranks as high among the virtues as it used to.)

Many tribal societies have elaborate rituals of gift-giving, always with the understanding that the recipient must repay. Often the repayment has to be superior to the original gift. Sometimes this escalation rises to such heights that people try at all costs to avoid receiving the gift, or try to pay it back immediately in order to be free of any obligation.

In the Western ethical tradition, too, gratitude and revenge have had a leading place. The investigation of justice undertaken in Plato's *Republic* gets under way by dissecting the popular view that justice consists in doing good to one's friends and harm to one's enemies. Cicero wrote that it is "the first demand of duty" that we do most for him that loves us most; "no duty is more imperative," he added, than that of proving one's gratitude. This is also the attitude to which Jesus referred in the Sermon on the Mount, when he said: "Ye have heard that it hath been said, thou shalt love thy neighbor and hate thine enemy." (Jesus proposed that we love our enemies instead of hating them, but even he found it necessary to hold out the prospect of a reward from God for doing more than publicans and sinners do.)

From our positive feelings for those who help us spring the bonds of friendship and the loyalty that we feel we owe to friends; from our negative feelings for those who do not reciprocate we get moral indignation and the desire to punish. If reciprocal altruism played a significant role in human evolution, an aversion to being cheated would be a distinct advantage. Humans have this aversion; indeed, we have it to such an extent that it often seems counterproductive. People who could not be induced to work an hour's overtime for ten dollars will spend an hour taking back defective goods worth five dollars. Nor is this lack of proportion unique to our culture. Anthropologists observing many different societies report bloody fights arising from apparently trivial causes. "It isn't the five dollars," we say in defense of our conduct, "it's the principle of the thing." No doubt the San "Bushmen" of the Kalahari say much the same when they fight over the distribution of the spoils of a hunt. But why do we care so much about the principle? One possible explanation is that while the cost of being cheated in a single incident may be very slight, over the long run constantly being cheated is much

more costly. Hence it is worth going to some lengths to identify cheaters and make a complete break with them.

Personal resentment becomes moral indignation when it is
shared by other members of a group and brought under a
general principle. Polybius, in the passage quoted above,
refers to others imagining themselves in the same situation,
with the same feeling of resentment, as the victim of ingrati-
tude. Because we can imagine ourselves in the position of
others, and we can formulate general rules which deal with
these cases, our personal feelings of resentment may solidify
into a group code, with socially accepted standards of what
constitutes adequate return for a service, and what should be
done to those who cheat. Though vengeance in tribal societies is often left up to the injured party and his or her kin,
there are obvious disadvantages in this system, since both
sides will often see themselves as having been wronged, and
the feud may continue to everyone's loss. To avoid this, in
most societies blood feuds have been replaced by settled
community procedures for hearing evidence in disputes, and
pronouncing an authoritative verdict that all parties must
obey.

Reciprocal altruism may be especially important within a
group of beings who can reason and communicate as humans
can, for then it can spread from a bilateral to a multilateral
relationship. If I help you, but you do not help me, I can of
course cease to help you in the future. If I can talk, however,
I can do more. I can tell everyone else in the group what sort
of a person you are. They may then also be less likely to help
you in future. Conversely, the fact that someone is a reliable
reciprocator may also become generally known, and make
others readier to help that person. "Having a reputation" is

only meaningful among creatures who communicate in a sophisticated manner; but when it develops, it immensely increases the usefulness of reciprocal altruism. If I have once
saved a person from drowning, and am later in need of rescue
myself, I will be lucky indeed if the very person I rescued is
within earshot. So if my heroic deed is known only to the
person I saved, it is unlikely to have future benefits. If, on the
other hand, my saving someone increases the likelihood of
any member of the community coming to my assistance, the
chances of my altruism redounding to my own advantage are
much better.

That the practice of reciprocal altruism should be the
source of many of our attitudes of moral approval and disapproval, including our ideas of fairness, cheating, gratitude,
and retribution, would be easier to accept if it were not that
this explanation seems to put these attitudes and ideas on too
self-interested a footing. Reciprocal altruism seems not really
altruism at all; it could more accurately be described as enlightened self-interest. One might be a fully reciprocating
partner in this practice without having the slightest concern
for the welfare of the person one helps. Concern for one's
own interests, plus the knowledge that exchanges of assistance are likely to be in the long-term interests of both partners, is all that is needed. Our moral attitudes, however, demand something very different. If I am drowning in a raging
surf and a stranger plunges in and rescues me, I shall be very
grateful; but my gratitude will diminish if I learn that my rescuer first calculated the probability of receiving a sizable reward for saving my life, and took the plunge only because the
prospects for the reward looked good. Nor is it only gratitude
that diminishes when self-interested motives are revealed;
moral approval is always warmest for acts which show either

spontaneous concern for the welfare of others or else a conscientious desire to do what is right. Proof that an action we have praised had a self-interested motive almost always leads us to withdraw or qualify our praise.

Early in the previous chapter, we accepted a definition of altruism in terms of behavior—"altruistic behavior is behavior which benefits others at some cost to oneself"—without inquiring into motivation. Now we must note that when people talk of altruism they are normally thinking not simply of behavior but also of motivation. To be faithful to the generally accepted meaning of the term, we should redefine altruistic behavior as behavior which benefits others at some initial cost to oneself, and is motivated by the desire to benefit others. To what extent human beings are altruistically motivated is a question I shall consider in a later chapter. Meanwhile we should note that according to the common meaning of the term, which I shall use from now on, an act may in fact benefit me in the long run, and yet—perhaps because I didn't foresee that the act would redound to my advantage—still be altruistic because my intention was to benefit someone else.

Robert Trivers has offered a sociobiological explanation for our moral preference for altruistic motivation. People who are altruistically motivated will make more reliable partners than those motivated by self-interest. After all, one day the calculations of self-interest may turn out differently. Looking at the shabby clothes I have left on the beach, a self-interested potential rescuer may decide that the prospects of a sizable reward are dim. In an exchange in which cheating is difficult to detect, a self-interested partner is more likely to cheat than a partner with real concern for my welfare. Evolution would therefore favor people who could distinguish self-interested from altruistic motivation in others, and then select only the altruistic as beneficiaries of their gifts or services.

Psychologists have experimented with the circumstances that lead people to behave altruistically, and their results show that we are more ready to act altruistically toward those we regard as genuinely altruistic than to those we think have ulterior motives for their apparently altruistic acts. As one review of the literature concludes: "When the legitimacy of the apparent altruism is questioned, reciprocity is less likely to prevail." Another experiment proved something most of us know from our own attitudes: we find genuine altruism a more attractive character trait than a pretense of altruism covering self-interested motives.

Here an intriguing and important point emerges; if there are advantages in being a partner in a reciprocal exchange, and if one is more likely to be selected as a partner if one has genuine concern for others, there is an evolutionary advantage in having genuine concern for others. (This assumes, of course, that potential partners can see through a pretense of altruism by those who are really self-interested—something that is not always easy, but which we spend a lot of time trying to do, and often can do. Evolutionary theory would predict that we would get better at detecting pretense, but at the same time the performance of the pretenders would improve, so the task would never become a simple one.)

This conclusion is highly significant for understanding ethics, because it cuts across the tendency of sociobiological reasoning to explain behavior in terms of self-interest or the interests of one's kin. Properly understood, sociobiology does not imply that behavior is actually motivated by the desire to further one's own interests or those of one's kin. Sociobiology says nothing about motivation, for it remains on the level of the objective consequences of types of behavior. That a piece of behavior in fact benefits oneself does not mean that the behavior is motivated by self-interest, for one might be quite unaware of the benefits to oneself the behavior will bring. Nevertheless, it is a common assumption that sociobiology implies that we are motivated by self-interest, not by genuine altruism. This assumption gains credibility from some of the things sociobiologists write. We can now see that sociobiology itself can explain the existence of genuinely altruistic motivation. The implications of this I shall take up in a later chapter, but it may be useful to make the underlying mechanism more explicit. This can be done by reference to a puzzle known as the Prisoner's Dilemma.

In the cells of the Ruritanian secret police are two political prisoners. The police are trying to persuade them to confess to membership in an illegal opposition party. The prisoners know that if neither of them confesses, the police will not be able to make the charge stick, but they will be interrogated in the cells for another three months before the police give up and let them go. If one of them confesses, implicating the other, the one who confesses will be released immediately but the other will be sentenced to eight years in jail. If both of them confess, their helpfulness will be taken into account and they will get five years in jail. Since the prisoners are interrogated separately, neither can know if the other has confessed or not.

The dilemma is, of course, whether to confess. The point of the story is that circumstances have been so arranged that if either prisoner reasons from the point of view of self-interest, she will find it to her advantage to confess; whereas taking the interests of the two prisoners together, it is obviously in their interests if neither confesses. Thus the first prisoner's self-interested calculations go like this: "If the other prisoner confesses, it will be better for me if I have also confessed, for then I will get five years instead of eight; and if the other prisoner does not confess, it will still be better for me if I confess, for then I will be released immediately, instead of being interrogated for another three months. Since we are interrogated separately, whether the other prisoner confesses has nothing to do with whether I confess—our choices are entirely independent of each other. So whatever happens, it will be better for me if I confess." The second prisoner's self-interested reasoning will, of course, follow exactly the same route as the first prisoner's, and will come to the same conclusion. As a result, both prisoners, if self-interested, will confess, and both will spend the next five years in prison. There was a way for them both to be out in three months, but because they were locked into purely self-interested calculations, they could not take that route.

What would have to be changed in our assumptions about the prisoners to make it rational for them both to refuse to confess? One way of achieving this would be for the prisoners to make an agreement that would bind them both to silence. But how could each prisoner be confident that the other would keep the agreement? If one prisoner breaks the agreement, the other will be in prison for a long time, unable to punish the cheater in any way. So each prisoner will reason: "If the other one breaks the agreement, it will be better for me if I break it too; and if the other one keeps the agreement, I will still be better off if I break it. So I will break the agreement."

Without sanctions to back it up, an agreement is unable to bring two self-interested individuals to the outcome that is best for both of them, taking their interests together. What has to be changed to reach this result is the assumption that the prisoners are motivated by self-interest alone. If, for instance, they are altruistic to the extent of caring as much for the interests of their fellow prisoner as they care for their own interests, they will reason thus: "If the other prisoner does not confess it will be better for us both if I do not confess, for then between us we will be in prison for a total of six months, whereas if I do confess the total will be eight years; and if the other prisoner does confess it will still be better if I do not confess, for then the total served will be eight years, instead of ten. So whatever happens, taking our interests together, it will be better if I don't confess." A pair of altruistic prisoners will therefore come out of this situation better than a pair of self-interested prisoners, *even from the point of view of self-interest.*

Altruistic motivation is not the only way to achieve a happier solution. Another possibility is that the prisoners are conscientious, regarding it as morally wrong to inform on a fellow prisoner; or if they are able to make an agreement, they might believe they have a duty to keep their promises. In either case, each will be able to rely on the other not confessing and they will be free in three months.

The Prisoner's Dilemma shows that, paradoxical as it may seem, we will sometimes be better off if we are not self-interested. Two or more people motivated by self-interest alone may not be able to promote their interests as well as they could if they were more altruistic or more conscientious.

The Prisoner's Dilemma explains why there could be an evolutionary advantage in being genuinely altruistic instead of making reciprocal exchanges on the basis of calculated self-interest. Prisons and confessions may not have played a substantial role in early human evolution, but other forms of cooperation surely did. Suppose two early humans are attacked by a sabertooth cat. If both flee, one will be picked off by the cat; if both stand their ground, there is a very good chance that they can fight the cat off; if one flees and the other stands and fights, the fugitive will escape and the fighter will be killed. Here the odds are sufficiently like those in the Prisoner's Dilemma to produce a similar result. From a self-interested point of view, if your partner flees your chances of survival are better if you flee too (you have a 50 percent chance rather than none at all) and if your partner stands and fights you still do better to run (you are sure of escape if you flee, whereas it is only probable, not certain, that together you and your partner can overcome the cat). So two purely self-interested early humans would flee, and one of them would die. Two early humans who cared for each other, however, would stand and fight, and most likely neither would die. Let us say, just to be able to put a figure on it, that two humans cooperating can defeat a sabertooth cat on nine out of every ten occasions and on the tenth occasion the cat kills one of them. Let us also say that when a sabertooth cat pursues two fleeing humans it always catches one of them, and which one it catches is entirely random, since differences in human running speed are negligible in comparison to the speed of the cat. Then one of a pair of purely self-interested humans would not, on average, last more than a single encounter with a sabertooth cat; but one of a pair of altruistic humans would on average survive ten such encounters.

If situations analogous to this imaginary sabertooth cat attack were common, early humans would do better hunting with altruistic comrades than with self-interested partners. Of course, an egoist who could find an altruist to go hunting with him would do better still; but altruists who could not detect—and refuse to assist—purely self-interested partners would be selected against. Evolution would therefore favor those who are genuinely altruistic to other genuine altruists, but are not altruistic to those who seek to take advantage of their altruism. We can add, again, that the same goal could be achieved if, instead of being altruistic, early humans were moved by something like a sense that it is wrong to desert a partner in the face of danger.

GROUP ALTRUISM AND HUMAN ETHICS

In the previous chapter we saw that most sociobiologists believe kin selection and reciprocity to have been more significant forces in evolution than group selection; nevertheless, we found some grounds for believing that group selection might have played a role. Whether or not group selection has been significant among non-human animals, when we look at human ethical systems the case for group selection is much stronger, although in view of the clear interest each society has in promoting devotion to the group, it is here even harder than in other cases to disentangle biological and cultural influences. What can be said for the biological side is that early humans lived in small groups, and these groups were at least sometimes reproductively isolated from each other by geography or mutual hostility; thus the conditions necessary for selection on a group basis existed. Cultural influences probably enhanced the tendency toward group altruism, by punishing those who put their own interests too far ahead of the interests of the group, and rewarding those who make sacrifices for the group.

In placing group altruism after kin and reciprocal altruism

we are following, once more, Sidgwick's hierarchy of the degrees of benevolence. He reports the morality of his day as placing the duty to be benevolent "to neighbours and to fellow-countrymen" immediately after the duty to be benevolent to friends, and before the duty to be benevolent to members of our own race. That we have a duty to assist the poor of our own neighborhood or nation before we assist the poor of another neighborhood or country is still a popular sentiment. It is part of the common belief that we should look after "our own" before we make efforts to help the starving overseas. I have already mentioned this view in connection with the ties of kinship, but once the obligations of kinship are fulfilled the boundaries of "our own" expand to the next-largest community with which we identify, whether this be a local or regional grouping, or an affiliation based not on living in the same area but on a shared characteristic like ethnic or class background, or religious belief. Beyond this priority of concern for the welfare of members of our particular group, there is also a loyalty to the group as a whole which is distinct from loyalty to individual members of the group. We tend to identify with a group, and see its fortunes as to some degree our fortunes. The distinction is easily seen at the national level, where "patriotism" describes a loyalty to one's nation that has little to do with helping individual fellow citizens.

Like kin altruism and reciprocal altruism, group altruism is a strong and pervasive feature of human life. When people live in small kinship groups, kin altruism and group altruism overlap; but the ethical codes of larger societies almost always contain elements of distinctively group altruism. It is very common for tribal societies to combine a high degree of altruism within the tribe with overt hostility to members of neighboring tribes. Similarly strong feelings of loyalty to one's group have been reported by anthropologists from many different cultures. The ancient Greeks particularly praised devotion to one's city-state, and we have seen how Plato thought that state loyalty should take precedence over family loyalty, at least among his Guardians. Cicero, in a characteristic piece of Roman rhetoric, wrote:

> Parents are dear; dear are children, relatives, friends; but one's native land embraces all our loves; and who that is true would hesitate to give his life for her, if by his death he could render her a service?

The persistence of group loyalty in modern times was only too clearly demonstrated by Hitler's success in arousing the nationalistic feelings of the German people, and Stalin's need to appeal to "Mother Russia" rather than the defense of Communism to rally the citizens of the Soviet Union to the war effort. In a less sinister way we can witness the appeal of group loyalty every weekend by watching the behavior of the crowds at football games.

Our ethical codes reflect our group feelings in two ways, corresponding to the difference between group altruism manifested as a preference for altruism directed toward individual members of one's own group, and group altruism manifested as loyalty to the group as a whole. We have seen the group bias of our ethics in respect to the first of these—the widespread and socially approved attitude that the obligation to assist people in other countries is much weaker than the obligation to assist our fellow citizens. The group bias of our ethics in respect to loyalty to the group as a whole shows itself in the high praise we give to patriotism.

Why is it that "my country, right or wrong!" can be taken

seriously? Why do we regard patriotism as a virtue at all? We disapprove of selfish behavior, but we encourage group self-ishness, and gild it with the name "patriotism." We erect statues to those who fought and died for our country, irre-spective of the merits of the war in which they fought. (One of the reasons why Robert E. Lee, leader of the Confederate Army in the Civil War, is such an admired figure in Ameri-can history is that he put his loyalty to his native Virginia above his publicly stated moral doubts about slavery.)

Patriotism has had its critics, among them many of the most enlightened and progressive thinkers. Diogenes the Cynic declared himself to be the citizen not of one country but of the whole world. Stoic philosophers like Seneca and Marcus Aurelius also argued that our loyalty should be to the world community, not to the state in which we happen to be born. Voltaire, Goethe, and Schiller espoused similar ideals of world, rather than national, citizenship. Yet patriotism has proved difficult to dislodge from its high place among the conventionally accepted virtues. The explanation for this could be that patriotism rests, at least in part, on a biological basis; but the explanation could also be cultural. Culture can itself be a factor in the evolutionary process, those cultures prevailing which enhance the group's prospect of survival. The prevalence of patriotism could easily be explained in this manner.

That cultural and biological factors interact is something that should be borne in mind throughout our discussion of the biological basis of ethics. Biological and cultural explanations of human behavior are not inconsistent unless, foolishly, we try to insist that one of these two is the sole cause of a com-plex piece of behavior. With certain exceptions, that is un-likely. Culture may intensify, soften, or perhaps under special conditions altogether suppress genetically based tendencies. Earlier in this chapter I referred to the extent to which practices based on racial and ethnic group feeling have been softened or eliminated by changes in attitudes. Here we have a clear example of something that may well have some biological basis—but also contains a strong cultural component—being altered by a cultural change. In a multiracial society, strong racial feelings are a disadvantage; strong patriotic feelings, however, are not.

One other cautionary note before I bring this chapter to a close: Up to this point our discussion has been purely descriptive. I have been speculating about the origins of human ethics. No ethical conclusions flow from these speculations. In particular, the suggestion that an aspect of human ethics is universal, or nearly so, in no way justifies that aspect of human ethics. Nor does the suggestion that a particular aspect of human ethics has a biological basis do anything to justify it. Because there is so much misunderstanding of the connection between biological theories about ethics and ethical conclusions themselves, the task of examining claims about this connection needs a chapter to itself.

3

FROM EVOLUTION TO ETHICS?

As long as we remain within the realm of science proper, we can never meet with a sentence of the type "Thou shalt not lie" ... Scientific statements of facts and relations ... cannot produce ethical directives.

—Albert Einstein, *Out of My Later Years*

... science may soon be in a position to investigate the very origin and meaning of human values, from which all ethical pronouncements and much of political practice flow.

—Edward O. Wilson, *On Human Nature*

THE TAKEOVER BID

We have now seen how—consistent with what we know of evolutionary theory—kin altruism, reciprocal altruism, and a limited amount of group altruism could have developed among the social animals from which we are descended; and could, quite naturally, have evolved into systems of ethics which in some respects resemble the ethical systems common among humans. Edward Wilson has claimed that sociobiological theories have great significance for human ethics. In this chapter I shall present, and then try to clarify and assess,

Wilson's claims. First, we need to see what these claims are. It is an indication of the importance Wilson places on them that they are the subject of the opening sentences of *Sociobiology:*

> Camus said that the only serious philosophical question is suicide. That is wrong even in the strict sense intended. The biologist, who is concerned with questions of physiology and evolutionary history, realizes that self-knowledge is constrained and shaped by the emotional control centers in the hypothalamus and limbic system of the brain. These centers flood our consciousness with all the emotions—hate, love, guilt, fear and others—that are consulted by ethical philosophers who wish to intuit the standards of good and evil. What, we are then compelled to ask, made the hypothalamus and limbic system? They evolved by natural selection. That simple biological statement must be pursued to explain ethics and ethical philosophers, if not epistemology and epistemologists, at all depths.

Although Wilson is clear that sociobiology should make a dramatic difference to ethics, he is regrettably less clear about exactly what difference it makes. For the next 560 pages of *Sociobiology* there is nothing to suggest how biology can explain ethics "at all depths" until, in the final chapter of the book, Wilson abruptly suggests that perhaps "the time has come for ethics to be removed temporarily from the hands of the philosophers and biologicized." As an illustration of the errors of non-biological ethical philosophy, Wilson takes John Rawls's well-known book, *A Theory of Justice.* Rawls derives principles of justice from the principles free and rational persons would choose if they were laying down the ground rules of a new association, starting from a position of equality. This conception of justice, Wilson says, may be "an ideal state for disembodied spirits," but it is "in no way explanatory or predictive with reference to human beings." Moreover, "it does not consider the ultimate ecological or genetic consequences of the rigorous prosecution of its conclusions."

So much for Rawls and, apparently, for ethical philosophers in general (no others are mentioned); but what, more positively, would a "biologicization" (if that is the noun form of Wilson's new verb) of ethics be like? Here is another hint:

> In the first chapter of this book I argued that ethical philosophers intuit the deontological canons of morality by consulting the emotive centers of their own hypothalamic-limbic systems. . . . Only by interpreting the activity of the emotive centers as a biological adaptation can the meaning of the canons be deciphered.

Then comes an outline sketch of an interpretation of the activity of the emotional centers as a biological adaptation. Some emotional activity will, Wilson suggests, be an outdated relic of earlier forms of tribal life. In other ways our emotions may be in the process of adapting to urban life. Impulses arising from altruistic genes established by group selection will be opposed by more egoistic impulses arising from genes favored by individual selection. Age and sex differences may cause further moral ambivalence. Evolution selects more strongly against altruism in young children than it does in older people who have already reproduced and who can therefore risk their lives without risking the survival of their genes. Females who must bear, and in the past had to feed, the infants have a stronger genetic interest in a durable relationship with a sexual partner than do males.

All this Wilson sees as leading to a theory of "innate moral pluralism" according to which no single set of moral standards is applicable either to all human populations or to all the different age and sex groups within each population. It is also supposed to show that "the requirement for an evolutionary approach to ethics is self-evident."

Wilson ended *Sociobiology* by saying that only when we can give a full neuronal explanation of the human brain, only when "the machinery can be torn down on paper at the level of the cell and put together again," will one be able to have a "genetically accurate and hence completely fair code of ethics." In more recent writings, however, he has been more explicit about the ethical recommendations to be derived from a sociobiological understanding of human nature.

In *On Human Nature* there is a discussion of the biology of sex which deals with three separate issues: differences in behavior between males and females; family life; and homosexuality. On the first issue, Wilson argues that there are genetic factors which lead girls to be, on average, more sociable and *No!* less physically venturesome and aggressive than boys. Similarly, he believes that our tendency to live in families has a biological basis in the care our children need during their long period of dependence. Sex is not only for reproduction: it serves to reinforce the bond between male and female, leading to cooperation in the raising of children. Even homosexuality, Wilson suggests, could be carried in our genes. Although homosexuals themselves have no offspring, they may be especially helpful to their kin. Relatives of homosexuals may thus survive and reproduce—and pass on the genes that occasionally result in homosexual behavior—at a higher rate than they would without the assistance of homosexual siblings or cousins. Wilson finds evidence for this kin selection theory of homosexuality in the tendency of homosexuals, both in tribal societies and in our own culture, to achieve higher status and influence than the average heterosexual with an otherwise similar background.

All these biological explanations of human behavior are controversial; but whether they are right or wrong is not the issue here. Our concern is with the ethical implications of these theories, not with the theories themselves. Wilson draws three ethical points from his biological theories. The first is a criticism of traditional "natural law" morality about sex. The idea that the primary role of sexual activity is reproduction is an error; hence attempts to condemn contraception and homosexuality on the grounds that they are "unnatural" are based on a mistake. Our nature is controlled by evolution, not by immutable divine command, so biologists rather than theologians are the real authorities on what is natural for us.

The second ethical point Wilson draws from his understanding of the biology of sex is that there will be costs, which we cannot yet measure, in implementing reforms which go against our biological tendencies. For instance, even though Wilson thinks there is a biological basis for the fact that males dominate many professions and cultural activities, he concedes that through quotas and a system of education which deliberately sets out to erase the differences between males and females, it would be possible to create a society in which there are as many females as males in any profession or activity. But while this would eliminate group prejudice and might result in a more harmonious and productive society, the amount of regulation it would require would jeopardize some personal freedoms, and it would prevent some individuals from realizing their full potential. Similarly with attempts to abolish families, or the opposite attempt to enforce the nuclear family as the norm, suppressing practices like homosexuality. Any attempt to set our culture against our nature costs us at least the time and energy required to inculcate and enforce a cultural standard which runs counter to our inherited tendencies. There may also be a greater cost: "Long-term defections from the innate censors and motivators of the brain can only produce," Wilson has elsewhere warned, "an ultimate dissatisfaction of the spirit and eventually social instability and massive losses in genetic fitness."

In the final chapter of *On Human Nature*, Wilson finds that sociobiology has ethical significance in yet a third way. He looks forward to the day when human biology "will fashion a biology of ethics, which will make possible the selection of a more deeply understood and enduring code of moral values." Although this is looking toward the future, Wilson thinks that already he can suggest three values that such a biology of ethics will support.

First comes "the cardinal value of the survival of human genes in the form of a common pool over generations." Wilson supports his claim that this is a cardinal value by noting that each individual is "an evanescent combination of genes drawn from this pool" and our own genes will soon be dissolved back into it. In the long-term view, our genes have come from millions of different ancestors, spread all over the world, and our descendants, in a few thousand years, will be similarly dispersed among millions of future human beings. So a detached view of evolution should lead us to consider the future of the entire human species, rather than just our own welfare and that of our kin or tribe.

Secondly, Wilson says, "a correct application of evolutionary theory also favors diversity in the gene pool as a cardinal value." The reason for this is that genius and other exceptional characteristics arise from rare combinations of diverse genes; reducing diversity may reduce the chance of these combinations occurring.

"Universal human rights," according to Wilson, "might properly be regarded as a third primary value." This is "because we are mammals." Mammals strive for reproductive success as individuals and cooperate only as a compromise to enjoy the benefits of group membership. The social insects, by contrast, reproduce as groups, and so if there were rational ants they would, Wilson imagines, regard individual freedom as intrinsically evil. Our mammalian nature is "the true reason for the universal rights movement."

So Wilson draws several different implications of sociobiology for ethics. Before I attempt to assess these claims, I shall try to clarify the different kinds of claims Wilson makes.

THE RELEVANCE OF BIOLOGY TO ETHICS: THREE POSSIBILITIES

Darwin's *Origin of Species* was published in November 1859. On January 4, 1860, Darwin wrote to a friend:

> I have received in a Manchester newspaper rather a good squib, showing that I have proved "might is right" and therefore that Napoleon is right, and every cheating tradesman is also right.

The Manchester reviewer was the first in a long line of writers who have drawn ethical implications from Darwin's theory of evolution. The line includes Herbert Spencer and the Social Darwinists; the anarchist Peter Kropotkin; and in our own century Julian Huxley and C. H. Waddington. Today

Edward Wilson is the most prominent representative of this line of thought.

Herbert Spencer is little read now. Philosophers do not regard him as a major thinker. Social Darwinism has long been in disrepute. Kropotkin appeals more to the romantic idealist in us than to the strictly scientific side of our intellect. Huxley and Waddington made little headway in their efforts to resurrect evolutionary ethics. Nevertheless, it would be wrong to brush aside Wilson's attempt to show the relevance of evolutionary theory to ethics just because the attempt has been made often before, and always in vain.

One reason why such dismissal would be premature is that sociobiologists could claim that earlier advocates of the evolutionary view failed only because we did not then understand enough about evolution; now that our knowledge of genes and their effects on behavior has developed, we are better prepared for the attempt to derive ethics from biology. Past failures do not preclude future success.

A more important reason for giving a hearing to Wilson's claims is that his is not an evolutionary ethics in the standard sense. He does not explicitly claim, as the Manchester reviewer took Darwin to be claiming, and many later proponents of evolutionary ethics really have claimed, that the course of evolution is itself good, and that therefore whatever advances evolution is good. This claim is quite out of keeping with Darwin's theory of natural selection, which sees evolution not as a purposive movement toward some ultimate goal, but as blind natural forces selecting some random mutations rather than others. Darwin himself was well aware that the "progress" of evolution is not progress in any ethical sense. To guard against misinterpretations that might flow from picturing evolution as a ladder reaching ever upward, he wrote himself a note: "Never use the words *higher* and *lower.*" Darwin's great champion, T. H. Huxley, went even further, saying, in a famous lecture, "Evolution and Ethics":

> Let us understand, once for all, that the ethical progress of society depends, not on imitating the cosmic process, still less in running away from it, but in combating it.

Huxley may have overstated his case, but the overstatement is at least more plausible than the opposite view—to which Huxley's grandson, Julian, swung back—that evolution necessarily moves in an ethically desirable direction.

Wilson avoids elevating evolution itself into a supreme ethical value. That much is clear. What then does he think follows, for ethics, from our new knowledge of biology? Wilson's scattered statements on exactly how biology has the potential to transform ethics do not make up a single, unified position. There are three distinct ways in which Wilson suggests that scientific findings are relevant to ethics. I shall first state all three briefly, and then consider each in turn at greater length.

1. Science may produce new knowledge about the ultimate consequences of our actions. Wilson's objection that Rawls's conception of justice "does not consider the ultimate ecological or genetic consequences of the rigorous prosecution of its conclusions" implies that ethical principles proposed by philosophers who write in ignorance of evolutionary theory may bring about a situation they neither foresee nor want.

2. Science may undermine existing ethical beliefs. It can do this by showing that some ethical views—like the traditional objections to sex which cannot result in reproduction—are based on mistaken assumptions about

what is or is not "natural" for human beings. More broadly, science can undermine existing ethical beliefs by "deciphering the meaning of the canons of morality" through an interpretation of them as the outcome of biological adaptation. Seen in this light, the canons of morality cease to appear as absolute, self-evident, or divinely commanded moral rules.

3. Science may provide us with a new set of ethical premises or a reinterpretation of old ethical premises. This, presumably, is what Wilson has in mind when in *Sociobiology* he derives an "innate moral pluralism" from his evolutionary approach to ethics; and when, in *On Human Nature*, he refers to "the ethical premises inherent in man's biological nature," and looks forward to the time when a new biology of ethics will make it possible for us to select a better code of moral values.

ETHICAL THEORIES AND BIOLOGICAL CONSEQUENCES

Of the three ways in which Wilson seeks to bring science to bear on ethics, the first is the most straightforward. Traditionally facts have been regarded as the domain of science, values as the domain of ethics. What consequences our actions will have is a question of fact—often a very difficult one, since it requires us to make predictions about the future, but still a question of fact. For sociobiology to tell us something about the ultimate consequences of our actions would not threaten this traditional division of territory between science and ethics. If philosophers working in ethics take little notice of genetics or evolutionary theory, this is because the important philosophical questions—like "What is good?"—have to be answered *before* we can use information about the consequences of our actions in deciding what we ought to do. Since information about the consequences of our actions does not tell us which consequences to value, but only which action will or will not bring about the consequences we do value, most ethical theories simply incorporate new information about the consequences of our actions into our ethical decisions in a way which does not affect the fundamental theory of value itself.

Utilitarianism provides the most obvious example. In its simplest, classical form, utilitarianism is the theory that an act is right if and only if it does at least as much to increase happiness and reduce misery, for all those affected by it, as any possible alternative act. Thus the basic utilitarian principle taken by itself says nothing about whether peace is better than war, truth better than lying, or socialism better than capitalism. The utilitarian decides which institutions, policies, and actions are right by looking at the consequences in each case. New knowledge, however important, can only affect the utilitarian's estimate of what institutions, policies, or actions will maximize happiness. It cannot throw doubt on the principle of utility itself. (Try to imagine factual information that contradicts the view that happiness is the only thing that is intrinsically good, and misery the only thing intrinsically bad.*) Conceivably, a utilitarian who favored an

* Don't confuse "Happiness is the only thing that is intrinsically good" with *"Everyone thinks* happiness is the only thing that is intrinsically good." It is easy to produce facts that contradict the latter claim—but this no more contradicts the former than facts contradicting "No one thinks there are flying saucers" contradicts "There are no flying saucers."

intensive program of quotas and education to ensure that 50 percent of all high-status positions in our society went to females might, if she accepted Wilson's view of the costs of such a program, change her mind and accept a less radical target like equality of opportunity. Different information about the effects of inequality on the lives of women unable to realize their potential might lead a utilitarian who had been content with equality of opportunity to become a more radical feminist. In neither case does the utilitarian have to alter her underlying ethical theory.

Since Wilson specifically refers to the theory of justice proposed by John Rawls, we can take Rawls's theory as a second illustration of this point. Rawls argues (fallaciously, I believe, but that is not the issue here) that justice requires that we first insist on the most extensive total system of equal basic liberties compatible with similar liberties for all; and then allow inequalities in material goods only to the extent that doing so would improve the position of the worst-off group. Now what "ultimate ecological or genetic consequences" has Rawls failed to consider? I can see none. One remark about Rawls that Wilson makes in *On Human Nature*—that "Rawls would allow rigid social control to secure as close an approach as possible to the equal distribution of society's rewards"—suggests that Wilson may have in mind the costs of rigid social control, which he elsewhere describes as a loss of personal freedom and an ultimate dissatisfaction of the human spirit. The remark I have quoted indicates, however, a complete misunderstanding of Rawls's position, for Rawls insists that maximizing equal basic liberties takes absolute priority over the distribution of income. Even after the maximum liberty principle has been satisfied, Rawls never advocates an equal distribution of society's rewards; he proposes a distribution that maximizes the welfare of the worst-off. If this requires greater rewards for those with special abilities as an incentive for them to use their abilities for the benefit of the economy as a whole and the worst-off in particular, Rawls would accept highly unequal rewards. Thus Rawls would reject any rigid social controls which restricted everyone's liberty, and he would also reject anything which led to such dissatisfaction as to make the worst-off in our society still less happy than they would otherwise be.

The same reply can be made to the Social Darwinist objection—which seems to be hinted at by Wilson's reference to "genetic consequences"—that putting Rawls's theory into practice would interfere with the struggle for survival, allowing the "unfit" to reproduce, and thus bringing about the genetic deterioration of our community. Even if it somehow really could be shown that catastrophic genetic decline was a consequence of redistributing wealth, all that would follow for Rawls's conception of justice is that to the extent that a specific redistribution of wealth would lead to a catastrophic genetic decline, that redistribution would be prohibited by Rawls's theory. Catastrophic genetic decline would presumably make the worst-off in our society—along with everyone else—still worse off, and so anything that would bring about such a decline could not be required by Rawls's theory of justice.

For any other ethic based on consequences or goals a parallel line of argument would show that the ethic cannot be invalidated by new knowledge about the likely consequences of our actions or policies. What, though, of an ethical theory which emphasizes not goals or consequences, but moral rules or the preservation of absolute rights, irrespective of consequences? Kant's moral theory is often taken as an instance of this kind of view. Kant wrote that one should never tell a lie, not even if a would-be murderer comes to your door asking if his intended victim is hiding within. Robert Nozick's theory of absolute individual rights, presented in his book *Anarchy, State and Utopia,* provides a more recent example. Here, it may seem, new factual knowledge would have a more significant impact than it does on consequentialist theories. A deeper understanding of biology and evolution might show that adherence to, say, Kant's ethic of inflexible moral rules will lead to some genetic or ecological disaster. In a sense, this is true. New information could show that following a specific set of moral rules will lead to catastrophe; but it is precisely the nature of these absolutist moral theories that those who hold them remain unmoved by the consequences of following the moral rules. "Let justice be done, though the heavens fall!" is the attitude they take. That is, to my mind, a weakness in all absolutist theories. It is, however, a weakness well known to philosophers. We do not need new scientific knowledge to make us aware of the fact that absolutist theories may lead to disaster. Ordinary common-sense knowledge was enough to lead most philosophers to reject moral theories which pay no attention to consequences. The new insights of sociobiologists can only add to the stock of our knowledge of possible disasters.

So to sum up the impact on ethical theories of new scientific knowledge about the possible consequences of actions and policies: New information may mean that an ethical theory which pays attention to consequences turns out to require actions different from those we previously thought it required; but the core of a consequentialist theory will remain unaffected. On the other hand, a theory which pays no attention to consequences will not be affected at all, although new knowledge of the consequences of acting on such a theory might make such a theory even less plausible to those not already committed to it than it would otherwise have been.

My own view, which reflects my consequentialist position in ethics, is that all who think about ethical issues should draw their conclusions on the basis of the best information available. When well-grounded biological theories are relevant to an ethical decision, they should be taken into account. The particular moral judgments that we end up making may reflect these theories. For this reason it is perfectly true that philosophers, along with everyone else, should know something about the current state of biological theories of human nature. To ignore biology is to ignore one possible source of knowledge relevant to ethical decisions.

There is, however, no justification here for dramatic claims about explaining ethics "at all depths" or fashioning a biology of ethics which will do away with the need for ethical philosophers. Even if we should uncritically accept the sociobiological view of human nature in its entirety, the new facts we would have learned would affect ethics only at a relatively superficial level. The central question of ethics, the nature and justification of fundamental ethical values, would remain untouched.

DEBUNKING ETHICS

The second way in which Wilson thinks biology can transform ethics is by undermining existing ethical beliefs. Where an ethical belief is explicitly based on an assumption about what is natural for human beings, there is no difficulty in seeing how biology can be a tool of criticism. Strictly speaking, the impact of biology here is not to render the ethical belief untenable, but to destroy the original justification for that belief. Thus if the accumulation of evidence eventually confirms Wilson's suggestion that it is natural for a percentage of human beings to be homosexuals, opposition to homosexuality on the grounds that it is unnatural cannot be defended. People may still hold that homosexuality is wrong, but they will have to find some other reason for holding this belief, or else be content to put it forward as some kind of self-evident moral intuition.

So biological theories could have an important effect on those who make their ethical judgments on the basis of some theory of "natural law." This does not, however, amount to an important impact on ethics as a whole, since natural law systems of ethics are not widely held outside religious, and especially Roman Catholic, circles. Suppose homosexuality really were unnatural. Very few philosophers in the secular universities would think that one can validly argue from that fact to the conclusion that homosexuality is wrong. Obviously there are many things, from curing diseases to using saccharin, that are unnatural but not therefore wrong. Moreover, to argue that because something is unnatural it is wrong, is to argue from a fact to a value—a move which, for reasons I shall give in the following section, is invalid.

There are hints in Wilson's writings of a more important way in which biological theories may undermine ethical beliefs. If we come to see specific rules of ethics as biological adaptations resulting from our evolutionary history, we may cease to regard those ethical rules as morally absolute or self-evidently correct. Let us see how this can happen.

Every day we act in ways that reflect our ethical judgments. We may not think much about these judgments. If we do think about them, we are generally content to trace them back to some other judgment, about which we have fewer doubts (perhaps because it is a judgment widely accepted in the community). And there we let the matter rest. Almost all the thinking we do about ethics involves connecting one ethical judgment to another, more fundamental one. Even moral philosophers who develop theories about what we ought to do rarely press deeper. Some of them explicitly say that philosophy can do no more than systematize our moral intuitions. We can criticize one moral intuition on the basis of others, they say, but we cannot criticize all or most of our moral intuitions at once. Like trying to apply a lever to raise the world, the task is doomed because we have nowhere else to stand.

Precisely because science is outside ethics, the scientific study of the origin of our ethical judgment is a fulcrum on which we can rest our critical lever. In itself, science cannot compel us to abandon a principle—a fulcrum is not a force—but coupled with a commitment to rationality, it can provide leverage against basic ethical principles.

Science provides leverage against some ethical principles when it helps us understand why we hold our ethical principles. What we take as an untouchable moral intuition may be no more than a relic of our evolutionary history. Wilson has said: "The Achilles' heel of the intuitionist position is that it relies on the emotive judgment of the brain as though that organ must be treated as a black box." This is not entirely fair to intuitionists, who do try to distinguish between genuine and apparent intuitions, rejecting those they see as the products of special self-interested or cultural biases. Still, Wilson has a point. Discovering biological origins for our intuitions should make us skeptical about thinking of them as self-evident moral axioms.

Consider, as an example, the preference for "our own" which leads us to pay less attention to the sufferings of those outside our community than to those inside it. As we have seen, many think it right and proper to give priority to those closer to us; this was a principle of popular morality in Sidgwick's time, as it is in ours, and no doubt was throughout most of human history. Without a biological explanation of the prevalence of some such principle, we might take its near-universal acceptance as evidence that our obligations to our family are based on a self-evident moral truth. Once we understand the principle as an expression of kin selection, that belief loses credibility.

This is why the effect of a demonstration that some form of behavior has a biological basis can be the opposite of what those who try to deduce ethical principles from biology usually claim. Far from justifying principles that are shown to be "natural," a biological explanation is often a way of debunking the lofty status of what seemed a self-evident moral law. We must think again about the reasons for accepting those principles for which a biological explanation can be given.

It is not only biological explanations which have the effect of debunking accepted ethical principles. To complete the process, we would have to explore the history of the ethical beliefs of our own particular society. Then we would find relics of our cultural history to place alongside the relics of our evolutionary history. For instance, the Western principle of the sanctity of human life—a principle which is unique in the sharpness with which it separates the wrongness of taking the life of any human being, no matter how severely defective, from the wrongness of taking the life of any non-human animal, no matter how intelligent—can, as I have argued elsewhere, be explained as the legacy of the Judeo-Christian world view, in which humans, but not animals, are made in the image of God and have immortal souls. For those of us who do not accept the authority of the Judeo-Christian religions, this explanation should lead to a critical re-examination of our belief in the sanctity of all and only human life.

One problem with this debunking effect of biological and cultural explanations is: Where does it stop? If all our ethical principles can be shown to be relics of our evolutionary or cultural history, are they all equally discredited?

Wilson's answer might be that explanations of ethics discredit only those ethical principles they show to be relics of an earlier stage of our history, better suited perhaps to a tribal society than to modern urban life. Other ethical principles will be shown to be biological adaptations which remain well suited to the contemporary human situation. Those principles are justified by evolutionary theory. They will be the principles we retain.

If this were Wilson's answer, it would be an attempt to use biology to justify ethical principles. This is the last and potentially most significant of the three ways in which Wilson suggests biology can be relevant to ethics.

A BIOLOGICAL BASIS FOR
ULTIMATE VALUES?

Is it possible that science should lead, not merely to information relevant to the application of ultimate ethical values, but to the ultimate ethical values themselves? Could biologists discover "ethical premises inherent in man's biological nature"?

Asking these questions brings us squarely up against what is probably the best-known tenet of modern moral philosophy: the doctrine that there is an unbridgeable gulf between facts and values, between descriptions of what *is* and prescriptions of what *ought* to be. The existence of this gulf was first pointed to by David Hume in the following celebrated passage from his *Treatise of Human Nature*, which appeared in 1739:

> I cannot forbear adding to these reasonings an observation which may, perhaps, be found of some importance. In every system of morality which I have hitherto met with, I have always remarked that the author proceeds for some time in the ordinary way of reasoning, and establishes the being of a god, or makes observations concerning human affairs; when of a sudden I am surprised to find that instead of the usual copulation of propositions *is* and *is not*, I meet with no proposition that is not connected with an *ought* or an *ought not*. This change is imperceptible, but is, however, of the last consequence. For as this *ought* or *ought not* expresses some new relation or affirmation, it is necessary that it should be observed and explained; and at the same time that a reason should be given for what seems altogether inconceivable, how this new relation can be a deduction from others which are entirely different from it.

Here is an example of the kind of reasoning Hume had in mind. First our author "makes observations concerning human affairs":

> Few persons realize the true consequences of the dissolving action of sexual reproduction and the corresponding unim-

portance of "lines" of descent. The DNA of an individual is made up of about equal contributions of all the ancestors in any given generation, and it will be divided about equally among all descendants at any future moment. . . . The individual is an evanescent combination of genes drawn from this pool, one whose hereditary material will soon be dissolved back into it.

From this our author draws a conclusion which is clearly evaluative: "the cardinal value of the survival of human genes in the form of a common pool." Though he does not put this conclusion in a sentence using "ought" or "ought not," it would follow from accepting the survival of human genes in the form of a common pool as a cardinal value, that we ought not to do anything which imperils this survival.

Our author is, of course, Edward Wilson and the passage is taken from *On Human Nature*. The question we must now ask is: Is there really an unbridgeable gulf between facts and values? If so, do Wilson and others who attempt to derive ethical premises from biology slide illegitimately over this gulf?

I believe that the answer to both these questions is affirmative. The error in moving from facts to values—also known as committing "the naturalistic fallacy," although strictly speaking this is the fallacy of defining values in terms of facts, rather than simply deducing values from facts—is not difficult to grasp. Values must provide us with reasons for action. It would be pointless to try to convince people that, say, the survival of the human gene pool is a cardinal value, unless once you had convinced your audiences of this, they regarded themselves as now having a reason for not endangering the survival of the human gene pool. If someone says: "I accept that the survival of the human gene pool is a cardinal value, but I remain as indifferent as I ever was to whether or not the human gene pool survives," we will reply that she does not *really* accept the survival of the human gene pool as a value at all. She may just be mouthing conventionally accepted views which she does not really endorse. If our holding certain values had no effect at all on what we chose to do, values would lose all their importance.

Now think about facts. Facts, by themselves, do not provide us with reasons for action. I need facts to make a sensible decision, but no amount of facts can make up my mind for me. Hence no amount of facts can compel me to accept any value, or any conclusion about what I ought to do.

Let us see how this works. Suppose I am thinking of giving five hundred dollars to an organization that is trying to assist the tribespeople of Valod, in eastern India. These people are poor and backward. They scrape a living from small, dry plots of land, using the most primitive farming methods. Their birth rate is high, but many of the children die in infancy. In bad years, there is starvation. In better years, malnutrition still stunts the growth of children, weakening their resistance to disease. A scheme of assistance has been carefully thought out by an aid organization which depends on voluntary donations from the public. It will provide irrigation and teach better farming methods, so that the Valod people can grow enough food to sustain themselves in good health. It will improve medical care and introduce family planning—which the Valod community supports—to ensure that the increased food supply is not outstripped by population growth. Though there are many imponderables in any assistance scheme, this is among the best that thoughtful, concerned human beings can devise.

There are other things I could do with five hundred dollars.

I could buy a lot of new clothes, or some stereo equipment. It could pay for a better holiday than my family and I could otherwise afford. Or I could just invest it—no doubt my children would one day find the accumulated sum useful. Of course, my needs and those of my family are not nearly as pressing as those of the tribespeople of Valod. We are already quite comfortably off. My money would do more to relieve suffering and increase happiness if it went to Valod; but it would increase my happiness, and that of my family, more if it went to me or my family.

These are some of the facts that cross my mind as I think about what to do with the money. Obviously, these facts do not settle the issue for me. They tell me what my options are. How I choose between these options will reflect my values. The facts do not tell me what I value. Do I value helping strangers in India above a little extra comfort or luxury for myself and my family? The gap between facts and values lies in the inability of the facts to dictate my choice.

Would more facts, or facts of a different kind, bridge the gap? What about those facts that sociobiologists think important: facts about the nature of human beings as biological organisms with a specific evolutionary history; facts about the genetic basis of altruism; and facts about the hypothalamus and limbic system of the brain, which produce our emotions?

As the sociobiologists say, we are evolved biological organisms and our brain and our emotions reflect the evolutionary adaptations that have enabled us to survive. Our values and ethical systems are the products of our evolved nature. Isn't it then possible that as our knowledge of biology and physiology advance, they should come to reveal ethical premises inherent in our biological nature, thus bridging the gap between facts and values?

The short answer to this is: "No, it is not possible." No science is ever going to discover ethical premises inherent in our biological nature, because ethical premises are not the kind of thing discovered by scientific investigation. We do not find our ethical premises in our biological nature, or under cabbages either. We choose them. To return to our example, suppose I am considering what to do with my five hundred dollars and someone tells me that since evolution selects only those genes that promote their own survival, my genes are not likely to incline me to altruistic acts toward strangers. I will, however, have genes that prompt me to look after myself and my immediate family. Suppose that I can see no reason to doubt the cogency of this view of evolution. How will this information affect my decision? Do I immediately say: "Oh well, it's just too bad for the Indians, but since my altruism is genetically limited to my kin, I'll use the money on a family holiday"? Of course not. Information about my genes does not settle the issue, because I, and not my genes, am making the decision.

The sociobiologist might retort that this is a willfully unscientific way of looking at ethics. Emphasizing the "I" who is making the decision implies that this "I" is a mysterious entity which, unlike every other living thing, is somehow not open to scientific scrutiny. Yet—the sociobiologist will say—we know that human beings are, along with all other living things, subject to the influence of their biological nature. We are now beginning to understand how the values that people have are affected by the activity of their hypothalamus and limbic system. Every ethical value is a value held by a biological organism, and every choice between values is a choice made by a biological organism. Therefore every value and every choice is, in principle, explicable by the biological sciences. To be sure (the cautious sociobiologist will add), our knowledge of the biological basis of human values is still incomplete; but the gaps are steadily being filled in. Given time, we will be able to "explain ethics and ethical philosophers at all depths." Then ethical decisions will no longer have to be made on the basis of emotive, subjective, and often arbitrary choices. They will have a solid scientific basis.

This retort still misconceives the nature of the distinction between facts and values. The issue is not one of *explaining* values. When Wilson objects to Rawls's theory of justice because it is "in no way explanatory or predictive with reference to human beings," he reveals a deep misunderstanding of what ethics and ethical philosophers are about. Neither Rawls nor other contemporary ethical philosophers are trying to explain or predict human actions. If they were, they would be scientists, not philosophers—and we would still need ethical philosophers to puzzle over what we ought to do. Admittedly, the distinction between science and philosophy was not always drawn where we draw it now, and some earlier philosophers, like David Hume, can be interpreted as seeking to explain human behavior. Immanuel Kant, on the other hand, was very clear that his notion of what is morally good—namely, action done solely for the sake of duty, without any thought of self-interest—is not to be discovered by experience or observation of human actions. He even concedes the possibility that "perhaps the world has hitherto never given an example" of such actions; yet, he believes, our reason may still command them. Perhaps here, as elsewhere, Kant was taking an extreme position: but that such a position is even possible shows how far *explaining* and *predicting* are from *prescribing* or *justifying*. The gap that separates them is the gap between facts and values.

Science seeks to explain. If successful, it enables us to predict how the world will be. Ethics consists, as Einstein put it, of directives. Directives offer advice or guidance on what to do. In themselves, facts have no direction. They are neutral about what we ought to do. To take an example well beloved by philosophers, the fact that the bull is charging does not, by itself, entail the recommendation: "Run!" It is only against the background of my presumed desire to live that the recommendation follows. If I intend to commit suicide in a manner that my insurance company will think an accident, no such recommendation applies.

Similarly—to return to Wilson's argument for the cardinal value of the human gene pool—the fact that my DNA was once distributed among millions of other humans, and will again in the future be distributed among millions of other humans, does not in itself entail that I should be especially concerned about the survival of the human gene pool. I may be entirely indifferent to the fate of my genes, however widely they may be dispersed; or, though I care deeply for my children and grandchildren because I know and love them, I may have no regard for more remote descendants whom I shall never live to meet. Thus the facts Wilson adduces do not automatically provide us with reasons for ensuring the survival of the human gene pool. This conclusion follows only if we assume, as Wilson appears to assume, that everyone is concerned about the future of his or her DNA. If this really were the case—it seems a strange assumption—then the survival of one's own DNA would be the cardinal value, and Wilson's argument would not be deducing a new value from facts at all; it would merely be showing what, in the light of the fact that our DNA will eventually be very widely dispersed, we should do to ensure the survival of our own DNA.

Putting this point in strictly logical terms may help to show where Wilson has gone wrong. Wilson argues:

Premise: Our genes came from a common pool and will return to a common pool.
Conclusion: Therefore we ought not to do anything which imperils the common gene pool.

Wilson's conclusion contains a value word, "ought," whereas there is no value word in his premise. This means that his conclusion cannot follow from that premise alone. We can accept the premise and reject the conclusion. To make his argument logically cogent it needs a value premise as well as the factual premise. For instance, the following argument would be sound:

First premise: Our genes came from a common pool and will return to a common pool.
Second premise: We ought not to do anything which imperils the long-term survival of our genes.
Conclusion: Therefore we ought not to do anything which imperils the common gene pool.

If we accepted the premises of this argument we would have to accept its conclusion. A value is brought in with the second premise, and so the conclusion legitimately contains the value word "ought." Now we can see, however, that the second, evaluative premise is open to question. Why ought we not to do anything which imperils the long-term survival of our genes? Wilson has not argued for this evaluative premise at all, and its truth is by no means self-evident.

Sociobiology offers an explanation of ethics. It provides a theory about why human societies have systems of ethics, and why the rules or "canons" of these ethical systems typically prohibit some actions, like killing members of one's tribe, and encourage others, like sharing food with kin. The sociobiological account of ethics is thus on the same level as anthropological or sociological accounts of ethics. As the preceding chapters indicate, I think the sociobiological explanation of the origin and development of ethics may well be right; but that is not the issue here. The issue is: What are the ethical implications of a scientific explanation of ethics? To answer this it may help to see the sociobiological explanation as a rival to an anthropological explanation, and not as a rival to philosophical theories about what we ought to do. The mistake made by sociobiologists who think that their explanation of ethics can tell us what we ought to do parallels that of anthropologists who thought that the diversity of morals between societies implies that people ought to follow the moral code of their own society. Explanations of what ethics is, whether anthropological or sociobiological, cannot tell me what I ought to do, because I am not bound to follow the conventions of my society, or to foster the survival of my genes.

The difference between offering a scientific explanation and making an ethical decision is one of standpoints. The standpoint of an observer is not the standpoint of a participant. As a scientist, I might have observed hundreds of people wondering whether to donate money to overseas aid organizations or to spend it on themselves and their families. I might have plausible theories that fit all the data on the choices people make in these situations. Yet if someone should raise with me the question of *my* giving money to overseas aid instead of spending it on my family, my theories and data will not tell me what to do. They are theories about what people do in these situations. They have nothing to say about what people should do.

Our ability to be a participant in a decision-making process, to reflect and to choose, is as much a fact about human nature as the effect of the limbic system on our emotions. To recognize our ability to choose as a plain fact is not to depart from a scientific viewpoint, or to believe in a mysterious entity known as "I" or "the self" or "the will" which makes its choices in a realm beyond all causal laws. Recognizing our ability to choose is compatible with holding that a complete causal account of our behavior is, in principle, possible. Some people say that if an observer could predict how we would choose, that would show our belief in our ability to choose was illusory; but this is a mistake. We would still be making genuine choices. Though our present knowledge of human behavior is very limited, we can often predict how a friend we know well will choose—without thinking, when our prediction turns out to be accurate, that our friend did not make a real choice.

The distinction between the standpoint of the observer and the standpoint of the participant is ineliminable. Further progress from the observer's standpoint will not make the participant's standpoint redundant. Even if my theories about the choice to give money to overseas aid organizations were so accurate and complete that I could predict how a person exactly like myself would choose, I would *still* have to choose. Moreover, it is a curious and significant truth that my choice might be the opposite of what my theory predicts someone exactly like me would choose—*without the theory thereby being refuted.* This apparent paradox is explained by the fact that when I predict *my* choice I cannot help learning of the prediction before I make my decision; when I predict someone else's choice I can keep that person from knowing of my prediction. In my own case, however, my knowledge of the prediction might affect my choice—for instance, I may resent the idea that my behavior is predictable, and so I may make the opposite decision in order to reassure myself that I am capable of genuine choice. Of course, an *outside* observer might predict this too. In other words, if my original prediction was that a person exactly like me would use the money to take his family on a holiday, an outside observer, knowing that I would want to reassure myself about the genuineness of my choice, would predict that I would give the money to help the Indians. As long as I never learned of this outside observer's prediction there would be nothing to upset its accuracy. If, however, I were told of the outside observer's prediction, that prediction would again be a factor in my choice, and I might be contrary once again. This too might be predicted, but if I learned of this new prediction . . . And so on indefinitely. The point should be clear by now: I cannot predict my own choice—nor can anyone else tell me of a prediction of my choice—in a way that relieves me of the necessity to decide. A prediction about my choice, no matter how well supported by scientific theory, is just one more fact to take into account in deciding. In all cases where we have a choice worth thinking about, I am capable of refuting any prediction of which I become aware. To pretend otherwise is to evade responsibility for one's own decisions.

ULTIMATE CHOICES

Neither evolutionary theory, nor biology, nor science as a whole, can provide the ultimate premises of ethics. Biological explanations of ethics can only perform the negative role of making us think again about moral intuitions which we take to be self-evident moral truths but can be explained in evolutionary terms. In keeping with the general viewpoint sketched in the preface to this book, we cannot look to religion for positive guidance either. We have to choose our ultimate ethical premises ourselves. Is there anything more to be said about the choice? Existentialist philosophers, who agree that we must choose our ethical premises, say there is not. Insisting that our freedom to choose means that we are not limited to merely "following nature"—"existence precedes essence," as they obscurely put this point—they propose that the choice of ultimate values is simply a commitment, a "leap of faith," which is beyond any rational assessment, and thus ultimately arbitrary. This conclusion smacks of desperation, for it implies that the leap of faith which one existentialist philosopher (Heidegger) made to Nazism is, in the end, no less justifiable than the leap which another existentialist (Sartre) made to resistance to the Nazis.

Desperate as it is, this kind of ethical subjectivism could turn out to be an accurate picture of the limits of ethical debate. One problem with the debunking effect of biological and cultural explanations of our ethical principles is that they threaten to discredit too much. If all our ethical beliefs can be accounted for by these means, they are all equally discredited; but we cannot do without all our ethical principles. We still have to decide what to do, and that decision calls upon our values. So we would have to resurrect some of the discredited ethical principles. On what basis are we to do that? Unless there is a rational component to ethics that we can use to defend at least one of our fundamental ethical principles, the free use of biological and cultural explanations would leave us in a state of deep moral subjectivism. Indeed, if there is no reason for preferring one ethical principle to another, we may as well hang on to the ones we had before. So what if they have now turned out not to be the self-evident moral absolutes we once thought them? In the absence of any rational basis for ethics, they are no worse than any other principle. If ultimate ethical premises cannot be derived from anywhere, they are starting points which we choose to accept or reject; and if there is no basis for making this choice, ethics ultimately rests on subjective judgments which are immune from criticism.

Wilson's statements about ethics leave him with no escape from ethical subjectivism, once the impossibility of deriving ethics from biology has been admitted; for Wilson regards all nonbiological ethics as a matter of emotion. He says that "ethical philosophers who wish to intuit the standards of good and evil" are really just consulting their "emotional control centers." If ethical judgments were nothing but the outflow of our emotional control centers, it would be as inappropriate to criticize ethical judgment as it is to criticize gastronomic preferences. Endorsing capital punishment would be as much an expression of our feelings as taking our tea with lemon, rather than milk.

There is an alternative to regarding ethics as no more than the outpouring of our emotions. Ethical judgments may have a rational component. In his zeal to take over ethics, Wilson overlooks this position, held by Socrates, Plato, Aristotle, the Stoics, Aquinas, Kant, Sidgwick, and many other philosophers. The debate over the roles played by reason and emotion in ethics has been at the center of Western ethical philosophy since its beginnings in ancient Greece; yet Wilson assumes without argument that reason has no significant part to play in ethics.

The fact that we choose our ethical premises does not in itself imply that the choice is arbitrary. Before we accept that it is, we should question the assumption that reason plays no role in ultimate ethical choices. If this assumption is false—if, consistent with biological and evolutionary theory, reason is an important factor in the development of ethics—the rejection of attempts to derive ethical premises from evolution will not mean that ethics is just a matter of taste or subjective feelings. Though we must choose our ethical premises, we may be able to choose rationally.

The next two chapters explore the role of reason in the development of human ethics. Though nothing I shall say will be overtly contrary to our knowledge that we are biological, evolved beings, in Chapter 4 this knowledge will temporarily recede into the background, because our capacity to reason now takes the center of the stage. Those eager to condemn the emphasis on reason as typical of a philosopher's idealistic picture of human nature are asked to hold their fire: the admittedly one-sided account to be presented in Chapter 4 will be filled out in Chapters 5 and 6, in which we shall return from the rarefied heights of pure reason to look at ethics for beings capable of reasoning and at the same time the product of the natural selection of genes.

4

REASON

The moral sense perhaps affords the best and highest distinction between man and the lower animals; but I need say nothing on this head, as I have so lately endeavoured to shew that the social instincts—the prime principle of man's moral constitution—with the aid of active intellectual powers and the effects of habit, naturally lead to the golden rule, "As ye would that men should do to you, do ye to them likewise"; and this lies at the foundation of morality.

—Charles Darwin, *The Descent of Man*

. . . and if I say again that daily to discourse about virtue, and of those other things about which you hear me examining myself and others, is the greatest good of man, and that the unexamined life is not worth living, you are still less likely to believe me. Yet I say what is true, although a thing of which it is hard for me to persuade you.

—Socrates, in Plato's *Apology*

The blind progress of evolution has thrown up several species capable of reasoning; but the reasoning powers of normal human beings far exceed those of any other species. This is not to say that humans always do reason well, but that they are capable of reasoning well. How has this capacity affected the development of ethics?

THE NATURE OF REASON

The capacity to reason is a special sort of capacity because it can lead us to places we did not expect to go. This distinguishes it from, say, the ability to type. As I work on the draft of this chapter, I am using both my capacity to reason and my ability to type. My ability to type produces the results I expect—that is, the words I choose to convey my thoughts appear on the paper in my typewriter, more or less as I wanted them to. My capacity to reason, on the other hand, has less predictable consequences. Sometimes an argument that appeared sound turns out to be fallacious. I may have to drop a position I formerly held, even abandon a project I find I cannot complete. Matters can also take a brighter turn: I may see a connection between two points that I had overlooked before. I may become persuaded of something that I did not previously believe. Beginning to reason is like stepping onto an escalator that leads upward and out of sight. Once we take the first step, the distance to be traveled is independent of our will and we cannot know in advance where we shall end.

Perhaps the most famous example of this process is Hobbes's discovery of the Euclidean method of reasoning, a discovery that deeply influenced his approach to philosophy. The story is that Hobbes was browsing in a private library when he chanced upon a copy of Euclid's *Elements of Geometry* which lay open at the 47th theorem. On reading the conclusion, he swore that it was impossible. So he read the proof, which was based on a previously proved theorem, which he then also had to read, and this referred him back to another, and so on until he was at last convinced that the theorem he had doubted really did follow from axioms he could not reject. (Thereafter Hobbes tried to apply a similar standard of demonstrative reasoning in his own work; but what Euclid had done for geometry proved more difficult to apply to political philosophy.)

Another example: The capacity to count must have emerged at an early point in human development. Some non-human animals can count. It is said that if four hunters go into a thicket and only three come out, baboons will keep away, for they know that someone is still there; but if a larger number go in and all but one come out, the baboons are fooled. Baboons can't count very well; still, if this report is true, the ability to count has practical value for them and baboons who can count a little may sometimes survive when less gifted baboons perish. The ability to count must have conferred a similar advantage on our own ancestors. Every human tribe uses numbers. Long before writing developed, people made permanent records of their counting by cutting notches in a stick or stringing shells on twine. They had no idea that they had stepped onto an escalator of reasoning that leads by strictly logical steps to square roots, prime numbers, and the differential calculus. Yet so it was: the place and time of advances in mathematics depend on a host of historical factors, but the advances themselves are logical consequences of using numbers. If we start with counting we will sooner or later learn to add and subtract; and then one day when we want to share out a haul of nuts among several people, we shall begin to divide; and if the division does not work out exactly we shall, in time, discover fractions. Measuring plots of land can lead to geometry, and if, like Pythagoras, we try to calculate the length of the hypotenuse of a right-angled triangle, we are led by means of square roots to those mysterious things, irrational numbers (like the number which, multiplied by itself, makes two). This is still not far along toward higher mathematics, but it is enough to show the nature of the progression which takes us from the most elementary practical operations to a level of thought removed from the physicial needs of ordinary people. No doubt human beings are able to reason and count because in the evolutionary struggle for survival, a set of genes which included the genes leading to these abilities was more likely to survive than a set of genes which did not; but once these abilities had emerged, the development of mathematics is explicable, not only in terms of genes, but also in terms of the inherent logic of the concepts of number.

Could there be a parallel between this account of the development of mathematics and the development of ethics? Can our theories about the origin of altruism be blended with this view of reason to form a plausible account of the origin and nature of ethics? What follows is an attempt to do just that. Like any theory of the origin and nature of ethics, it contains speculations for which hard evidence is lacking. We are dealing with a stage of human development for which there are no historical records, and ideas leave no fossils. Nevertheless, the account I shall give is internally coherent and fits the evidence available; which is more than can be said for purely biological accounts that ignore the inherent logic of ethical thinking.

THE FIRST STEP

As we have seen, many non-human animals assist their own kin, or refrain from harming them. In some species this is true of unrelated animals as well. So the first steps toward ethics, like the first steps toward mathematics, were taken by our pre-human ancestors. Ethics starts with social animals prompted by their genes to help, and to refrain from injuring, selected other animals. On this base we must now superimpose the capacity to reason.

Imagine a group of early humans interacting on the basis of kin and reciprocal altruism. From kin altruism we get a strongly developed tendency for parents to provide for and defend their offspring. Brothers and sisters would help each other to a lesser extent. Nieces, nephews, and cousins receive preferential treatment over non-relatives, to a degree that diminishes with the distance of the relationship. Nevertheless, in a small, cohesive community, all members might be related to some degree, and so kin altruism, supplemented perhaps by a weaker tendency toward group altruism, would lead to a general readiness to help other members of one's own group, rather than other humans belonging to a different group. Reciprocal altruism adds a different and to some extent conflicting current: gratitude to those who have helped in the past leads to friendship and mutual aid which is not based on close kinship; and hostility to "cheats" who do not reciprocate would counteract any general benevolence toward all members of the group.

So far, this description could hold for non-human social animals as well as for human beings, although reciprocal altruism would be likely only among intelligent animals living in relatively stable groups. Gradually, as we evolved from our pre-human ancestors, our brains grew and we began to reason to a degree no other animal had achieved. We became better able to communicate with our fellows. Our language developed to the point at which it enabled us to refer to indefinitely many events, past, present, or future. We became more aware of ourselves as beings existing over time, with a past and a future, and more conscious of the patterns of our social life. We could reflect, and we could choose on the basis of our reflections. All this gave us, of course, tremendous advantages in the evolutionary competition for survival; but it also brought with it something which has not, so far as we can tell, occurred in any non-human society: the transformation of our evolved, genetically-based social practices into a system of rules and precepts guiding our conduct toward one another, supported by widely shared judgments of approval for those who do as the rules and precepts require, and disapproval for those who do not. Thus we arrived at a system of ethics or morality.

The transformation must have been a gradual one, over hundreds of thousands of years. The difference made by reason in this transformation is the difference between responding with a friendly lick or an intimidating growl when another member of the group does or does not repay favors, and responding with an approving or a condemnatory judgment. Stating the difference this way leaves open how big a difference it is—some may think that ethical judgments really are no more than refined friendly licks and intimidating growls. One difference, though, is apparent: growls and licks leave little to be discussed; ethical judgments leave a lot. To judge, beings have to be capable of thinking and of defending the judgments they make. Once beings can think and talk, once they can challenge each other, and ask, "Why did you do that?" their growls and licks are evolving into ethical judgments.

The notion of a judgment carries with it the notion of a standard or a basis of comparison, against which the judging is done. Because a judgment can be challenged, it is not limited to a specific occasion, as is a growl or lick. A dog may growl at one stranger and wag her tail at another without having to justify the apparent discrimination; but a human being cannot so easily get away with different ethical judgments in apparently identical situations. If someone tells us that she may take the nuts another member of the tribe has gathered, but no one may take her nuts, she can be asked why the two cases are different. To answer, she must give a reason. Not just any reason, either. In a dispute between members of a cohesive group of reasoning beings, the demand for a reason is a demand for a justification that can be accepted by the group as a whole. Thus the reason offered must be disinterested, at least to the extent of being equally acceptable to all. As David Hume put it, a person offering a moral justification must "depart from his private and particular situation and must choose a point of view common to him with others; he must move some universal principle of the human frame and touch a string to which all mankind have an accord and symphony."

The requirement of disinterestedness means that a barefaced appeal to self-interest will not do. I cannot say that I may take nuts from others because it benefits me, whereas when others take nuts from me, I lose. If I hope to gain the assent of the group as a whole, I must at least give my case an impartial guise. I may say, for instance, that my prowess as a warrior entitles me to a bigger share of the nuts. This justification is impartial in the sense that it entails that anyone who equals my prowess as a warrior should get as many nuts; conversely, it entails that if my fighting skills fall off, I will be entitled to fewer nuts. Cynics may say that I support the claims of warriors only because I happen to be one; if I were a gatherer of nuts, I would support the claim of the nut gatherers. Be that as it may, the need to clothe my justification with a semblance of disinterestedness provides a significant foothold for the development of ethical reasoning, as we shall soon see.

The dispute over nuts is, of course, an artificial example. It imports the modern readiness to question things into a tribal context. Early ethical systems, like the ethical systems of many cultures today, were probably marked less by deliberate questioning than by habitual acceptance. The medium between animal altruism and modern ethics was a system of social customs. The customs of a society are an accumulation of its collective approvals and disapprovals. Our language still shows this: "ethics" comes from the Greek *ēthos*, which normally means "character" but can, in the plural, mean "manners," and is related to the Greek word for custom, *ethos*, which differs only in having a shorter *e*. Similarly "moral" comes from the Latin *mos* and *moralis*, which mean "custom" and are related to the term "mores," which we still use to refer to the customary manners of a society. In tribal societies, "It is not customary" often has the force that "It is wrong" has for us. As inherently public, customs are necessarily impartial between individuals, in form at least. They may oppress whole groups, like women, or the poor, but they do so in a way that the oppressed can—and often do—accept as proper. To be a victim of oppressive customs is very different from being a victim of personal malice.

To a modern questioning mind, "It is customary" is so weak a reason for doing something that the appeal to custom may seem the very opposite of an appeal to reason; nevertheless, the transformation of genetically based social behavior into social customs involved the first limited application of reason to what had hitherto been under the unchallenged control of our genes. For the idea of custom implies the capacity to see beyond particular events, to group what happens here and now with what happened some time ago. This is a capacity which, though probably not limited exclusively to human beings, is certainly more highly developed in humans than in other animals. The readiness with which we can bring particular events under a general rule may be the most important difference between human and animal ethics.

Following custom seems like blind and unthinking obedience to the habits of the past. So it is, if we look at customary morality as a whole, with the perspective of an outside observer who can compare various systems of ethics and ask: "Why follow *this* one?" But this perspective is only possible much later in the development of reasoning about ethics. The early participants in the development of ethics had little awareness of possibilities beyond the actual practices of their own tribe or society. Within that perspective, the detailed operation of a customary morality needs reasoning to settle whether a particular action falls within the bounds of custom, and to extend old customs to new situations. To see what a task this can be, one has only to look at the English Common Law, which is really no more than a vast refinement of customary standards. In Common Law, the justification for any decision is always, on the surface, a past decision or precedent. The whole system supposedly expounds the common wisdom of the English people, as embodied in their customs and past practices. Of course, where the precedents are ambiguous, judges may lean toward the verdict they themselves think best, but they still must justify their decision in terms of precedent. The process often involves days of legal argument before the judge, and the judicial verdict can be as carefully reasoned—within its own more limited framework of precedent—as any work of philosophy. It would be a mistake to dismiss the reasoning involved in the development of a set of customary rules as insignificant because it does not reach beyond the confines of what has or has not been done in the past.

REASON PROGRESSING

Reason breaking beyond the boundary of customary morality is classically embodied in the life and death of Socrates. In Plato's portrayal of him, Socrates spends his life wandering through the markets of Athens questioning the conventional morality of his day. His method is to lure someone into claiming that it is a simple matter to understand the nature of justice, or courage, or piety, or some other virtue; then Socrates probes the definition offered until it becomes clear that it is entirely inadequate. Thus he leads his audience into the form of discussion that bears his name, a Socratic dialogue in which two people set out to find, by means of rational inquiry, what the good man ought to do. For instance, against the received wisdom that justice consists in paying one's debts, Socrates puts the example of a friend who lent you a weapon and, having since become deranged, asks for it back. Conventional morality gives no clear answer to this dilemma; its original definition of justice has to be reformulated, and the argument is under way. Socrates himself never claims to know the answer—his wisdom consists, he says, in the fact that he knows that he knows nothing. Therefore he knows more than those who know nothing but think they know something. That is the starting point of his criticism of conventional morality.

We all know what happened to Socrates. He was condemned to death and made to take hemlock. Though the trial was instigated by his political enemies, the charge was appropriate: corruption of the youth of Athens. By the standards of customary morality Socrates *was* corrupting the youth, for customary morality cannot stand the scrutiny of rational inquiry which questions the customary standards themselves. "The unexamined life is not worth living," Socrates told those about to vote on his sentence. He did not expect to be believed, and he was not believed. Perhaps he is still not widely believed. Few of us examine our lives in the Socratic manner. On the other hand, Socrates' example is universally admired while the actions of those who condemned him now seem contemptible.

The customs of the fifth century B.C. have passed away. Socrates' criticism of them no longer threatens us. Reason endures. Though we may find it hard to shake ourselves free of the customs of our own time, as rational beings we are drawn to respect someone utterly committed to rational inquiry and a rational life. In this way, despite Socrates' personal fate—or perhaps because of it—reason has progressed.

One example does not take us very far, no matter how dramatically it portrays a larger process. We need to look more closely at the logical basis for reason's progress. We have seen how reasoning beings can turn genetically based practices into a system in which custom acquires moral force. Applying a customary system of morality involves reasoning, sometimes at a very high level, but reasoning within limits. The next stage is that the customs themselves are questioned, as Socrates questioned the accepted standards of his time. Reaching this stage requires no mere continuation of previous trends, but a leap into the unknown. It is impossible to predict at what point in the development of a system of customary morality a Socratic questioner will appear, although, as already suggested, the ability to take an outsider's perspective on one's own society may have something to do with it. From the outsider's point of view, the customs of my own society appear as one among a number of different possible systems. Thus they lose their sense of natural rightness and inevitability.

Certainly the Greeks had, by Socrates' time, become aware of the variations in customs between different people. They argued over what is natural, and what customary, in human behavior. The Greek historian Herodotus chronicled different practices from many lands. He tells a story of how Darius, King of Persia, once summoned Greeks before him and asked them how much he would have to pay them to eat their fathers' dead bodies. They refused to do it at any price. Then Darius brought in some Indians who ate the bodies of their parents, and asked them what would make them willing to burn their fathers' bodies. The Indians cried out that he should not mention so horrid an act. Herodotus himself did not follow up the point, other than to remark that each nation thinks its own customs best, and only a madman would ridicule customs. Herodotus may have been an early cultural relativist, believing that everyone ought to follow the customs of his own society. This is a common initial response to knowledge of the differences between customs, but it manifestly is not a logical implication of that knowledge, and provides no rational basis for action. It is Socrates' search for a rational basis for ethics, rather than the relativist response, that marks the line toward a stage of moral thought beyond customary morality.

Interestingly, Lawrence Kohlberg, an American psychologist who has conducted studies of the moral development of children, adolescents, and adults in several different cultures, has also found that individuals whose moral thought is grounded in the conventions of their society often go through a period of relativism in the process of moving beyond conventional thinking to an independently thought-out ethics. Extreme relativism may be the first response to the shock of realization that one's ethical views are grounded in custom, and that if one had been brought up in a different society with different customs, one would have different ethical views; but the students Kohlberg studied all eventually came to hold that morality has some rational basis and therefore is not, in the end, arbitrary.

So seeing the customs of one's own society as one among several possible systems is likely to lead to the appearance of a questioner like Socrates. Of course, those devoted to upholding the established way of life are often more successful in crushing dissent than the Athenian establishment was in dealing with Socrates. Insofar as the timing and success of the emergence of a questioning spirit is concerned, history is a chronicle of accidents. Nevertheless, if reasoning flourishes within the confines of customary morality, progress in the long run is not accidental. From time to time, outstanding thinkers will emerge who are troubled by the boundaries that custom places on their reasoning, for it is in the nature of reasoning that it dislikes notices saying "off limits." Reasoning is inherently expansionist. It seeks universal application. Unless crushed by countervailing forces, each new application will become part of the territory of reasoning bequeathed to future generations. Left to itself, reasoning will develop on a principle similar to biological evolution. For generation after generation, there may be no progress; then suddenly there is a mutation which is better adapted than the ordinary stock, and that mutation establishes itself and becomes the base level for further progress. Similarly, though generations may pass in which thinkers accept conventional limits unquestioningly, once the limits become the subject of rational inquiry and are found wanting, custom has to retreat and reasoning can operate within broader bounds, which then in turn will eventually be questioned.

THE RATIONAL BASIS

Asking questions is one thing: getting answers to them another. The nature of the reasoning process may make it likely that customary standards will in time be questioned; but how do we get any further than questioning? If ethics derives from practices which have a genetic rather than a rational basis, isn't rational inquiry into ethics bound to be in vain? Isn't it searching for a rational basis which simply does not exist? More than two thousand years after Socrates' death, we all admire his questioning spirit, but how far have we got with answering his questions?

Reasoning in ethics is not limited to the negative task of rejecting custom as a source of ethical authority. We can progress toward rational settlement of disputes over ethics by taking the element of disinterestedness inherent in the idea of justifying one's conduct to society as a whole, and extending this into the principle that to be ethical, a decision must give equal weight to the interests of all affected by it. This would require me, in making an ethical judgment, to take my decision from a totally impartial point of view, a point of view from which I disregard my knowledge of whether I gain or lose by the action I am contemplating. One way of arriving at such a decision—an idea first suggested by the American philosopher C. I. Lewis and recently revived by R. M. Hare, professor of moral philosophy at the University of Oxford—is to imagine myself living the lives of all affected by my decision, and then ask what decision I prefer.

To see how this would work, imagine that I have to decide whether to keep a dinner appointment with three friends or to visit my father, who has just phoned to say that he is ill and must stay in bed. If no one except my friends, my father, and I will be affected by what I do, I can arrive at an impartial decision by imagining myself in the position of my friends if I break the appointment, and of my father, if I fail to visit him. There are also my own preferences to take into account. Let us say that I will enjoy the dinner more than visiting my father; but my father's disappointment at my not visiting him will far exceed the regrets of my friends at my absence from the table. To decide impartially I must sum up the preferences for and against going to dinner with my friends, and those for and against visiting my father. Whatever action satisfies more preferences, adjusted according to the strength of the preferences, that is the action I ought to take.

Adding up preferences in this way is, of course, a matter of judgment, for preferences do not come with labels attached, indicating how strong they are. (By "strong" in this context I mean, roughly, the importance the preference has for the person whose preference it is.) By imagining ourselves in the position of others, however, and taking on their tastes and preferences, we can often arrive at a reasonably confident verdict about which action will satisfy more preferences. For instance, if I know my father and my friends well, I might be quite sure that the regrets of my friends, even when trebled to take account of the fact that there are three of them, and even when my own loss of enjoyment is added in, will not be as significant as the disappointment of my father. Imagining myself living, one by one, the lives of myself, my friends, and my father, I may know that I would prefer the series of lives in which I visit my father to the series of lives in which I go to dinner with my friends.

Adding and subtracting preferences in this manner—and taking nothing but preferences or interests into account—is one way of settling ethical disputes; but is it the only way? Are we compelled by reason to take this approach? This may be the most difficult, but also the most important, question we can ask about ethics. For *if* this is the only rational way of reaching ethical judgments, ethics has a rational basis. We would then have an understanding of how to go about resolving ethical disputes. This would not mean that we could actually reach agreement about every ethical issue—there are lots of factual issues on which we cannot reach agreement, and in predicting the consequences of our actions and in estimating the strength of the preferences affected by them, we would certainly differ—but at least we would have criteria by which we could test the soundness of our ethical choices.

Before one attempts to answer this question, there is one point about it which needs to be clarified. In talking now of a rational basis for ethics, or of reason compelling us toward an impartial point of view, I have in mind "us" in a collective sense, not each one of us individually. My concern here is with how we might choose a morality for the group. Hence the focus on what the collective reasoning process leads us to; what it is rational for any given individual to do is a different question which will be discussed in the next chapter.

So the question we are considering is: Does the element of

disinterestedness inherent in the idea of justifying one's conduct to society lead us to a stance from which we give equal weight to the interests of all? To answer this question, we need to consider what other stances, consistent with the requirement of disinterestedness, we might take.

At the opposite end of the spectrum from the principle of equal consideration for the interests of all is some kind of egoism. Egoism holds that I ought to do what will further my own interests, not bothering about the interests of others, unless it so happens that in helping them I help myself. For this to be a possible basis for a group ethic, it must be stated in a form which makes it clear that this is not just a principle aimed at benefiting me. (The ideal form of egoism for me, Peter Singer, would be one that holds that everyone ought to do whatever furthers *my* interests—but that is not likely to be acceptable to others.) So for egoism to be a possible basis of ethics, it must be a disinterested form of egoism which holds that not just I but *everyone* ought to do what is in his or her interests.

Is there any reason why we should accept, as the basis of our ethics, the principle that everyone ought to do what is in his or her interests? Following Adam Smith, economists often claim that through the competition of the marketplace the individual pursuit of self-interest leads to the greatest good of all, because to profit one must sell something that is better or cheaper than one's competitors' products. If we accept this, we might hold that everyone ought to do what is in his or her interests simply because that is the way to promote the interests of all. But if we come to an egoistic ethics on this basis, we are not really egoists at all. Our most basic value is the good of all, impartially considered, and we have adopted egoism only as a means of gaining this end. If the facts of human life really are as Adam Smith and his followers claim, this form of ethical egoism follows from giving equal consideration to everyone's preferences. (I doubt that the facts are as these defenders of the free market claim—but that is a separate issue.) So this argument for egoism is not after all a rival to the method of resolving ethical disputes by summing up the preferences of all those affected; if valid at all, it is valid as a means of satisfying the greatest total of preferences.

A different defense of disinterested egoism, one which would present it as a genuine alternative to considering equally the interests of all, would be the claim that it is right or reasonable for everyone to further his or her own interests, *irrespective of the consequences of this for others.* This position disavows any concern with the consequences of people individually pursuing their interests; it maintains that this is what people should do, whether that turns out to be good or bad for all of us. In this respect it is like non-consequentialist moral claims, for instance the claim that one ought never to tell a lie, whatever the consequences.

This is a *possible* ethical position. It is disinterested—its advocates would need to have as little concern for their own interests as they have for the interests of others—and it is not self-contradictory. Someone could adopt it and be unmoved by any objections. Yet why should any group of rational beings accept it? It is potentially disastrous. Why make the basis of our ethics a principle which is oblivious to the consequences its adoption will have on each and every member of our group?

An advocate of this form of egoism might reply: "There is a reason why you should adopt this form of egoism. You should adopt it simply because it is right for everyone to further his or her own interests."

What would a person who says this mean? It could be just an expression of the speaker's personal support of this principle. But if this support has no basis other than subjective preferences, it ought to count no more in a moral argument than any other preferences. That is, it should be taken into account, along with the preferences of others. It should not be allowed to settle the issue simply because it is a preference for a moral principle rather than a preference for, say, food. If moral judgments are only subjective preferences, it would seem we deal properly with them by weighing them impartially alongside other preferences of those affected by our decisions.

But perhaps the advocacy of egoism is not simply an expression of a subjective preference, but a claim that egoism is a true moral principle, irrespective of the consequences of adopting it. In support of this general view of moral principles—though not of egoism—one could quote Thomas Carlyle, who once wrote: "What have men to do with interests? There is a right way and a wrong way. That is all we need think about it."

Those who take this view of ethics assume that there is truth or falsity in ethics, independently of the preferences of living beings. They often say that while the interests of humans—and other animals—are short-lived and parochial, the laws of ethics are eternal and universal. The laws of ethics, they say, existed before there was life on our planet and will continue to exist when the sun has ceased to warm the earth.

The grain of truth in this view of ethics is that there is *something* in ethics which is eternal and universal, not dependent on the existence of human beings or other creatures with preferences. The process of reasoning we have been discussing is eternal and universal. That one's own interests are one among many sets of interests, no more important than the similar interests of others, is a conclusion that, in principle, any rational being can come to see.

Wherever there are rational, social beings, whether on earth or in some remote galaxy, we could expect their standards of conduct to tend toward impartiality, as ours have. (Though the constraints limiting this tendency could be much stronger or much weaker.) But this universal element of ethics is so abstract that although we may say that it "exists" whether or not there are humans or other creatures with preferences, without the existence of some beings with preferences, the universal element is meaningless. If there are no beings with interests, the requirement that we treat all interests equally is entirely empty. It exists only as a framework into which the deliberations of rational creatures with preferences fit, when there are such creatures. It does not exist as a moral law commanding particular actions.

When morality was thought of as a system of laws handed down from on high, it was natural to think of moral judgments as attempts to describe moral laws which exist independently of us. The reality behind moral judgments seemed to be the will of God. Perhaps the legacy of past belief in a divine legislator is responsible for our ready assumption that there is something "out there" which our ethical judgments reflect. Now, however, the existence of ethics can be explained as the product of evolution among long-lived social animals with the capacity to reason. Hence, the need for belief in laws of ethics existing independently of us disappears. And the more we think about what it could mean—outside of a religious framework—for there to be eternal moral truths existing independently of living creatures, the more mysterious it becomes. As the Oxford philosopher J. L. Mackie has said of the idea of objective values:

> If there were objective values, then they would be entities of a very strange sort, utterly different from anything else in the universe. Correspondingly, if we were aware of them, it would have to be by some special faculty of moral perception or intuition, utterly different from our ordinary ways of knowing everything else.

[handwritten margin note: More say yes.]

The reason why objective values would have to be entities of a strange sort can be grasped from our previous discussion of the distinction between facts and values. Values are inherently practical; to value something is to regard oneself as having a reason for promoting it. How can there be something in the universe, existing entirely independently of us and of our aims, desires, and interests, which provides us with reasons for acting in certain ways? To borrow again from Mackie: How can "to-be-pursuedness" or "not-to-be-doneness" be built into the very nature of things?

We were discussing ethical egoism. In challenging the notion of moral principles which are true independently of their consequences, I have used heavy artillery to swat a fly. "Everyone ought to further his interests, whatever the consequences" was not a very plausible moral principle anyway. But the philosophical artillery would have had to be brought to bear, sooner or later, for there are more attractive moral principles which, like this form of egoism, satisfy the requirement of disinterestedness without adding up the preferences of all those affected. The conventional moral rules fall into this category. "Do not lie," "Never take an innocent human life," "Keep your promises," "Do not commit adultery"—all imply the existence of ultimate moral standards independent of the interests of those affected.

Of course, some at least of these conventional moral rules could turn out to be sound principles of conduct—but not absolutely exceptionless rules—in the sense that they are the principles a society should adopt if it is seeking to maximize the interests of all, impartially considered. This would give them a dependent validity, like the dependent validity of egoism if, as the free market economists claim, we would all be better off if we were each devoted to the exclusive pursuit of our own interests. In Chapter 6 I shall return to this question of the principles a society should adopt in order to further the interests of all; at the moment my concern is to deny that the conventional moral rules are valid in their own right, irrespective of their good or bad effects. Only this stronger claim is a threat to the idea that impartial consideration of the interests of all is the rational basis of ethics.

My grounds for denying that it could be rational to accept the conventional moral rules as independently valid principles of ethics are identical to those on which I argued that we should reject the principle of egoism when this principle was presented as valid irrespective of its consequences for us all. If the moral rules are not to be recommended to the group on the grounds of their good consequences for the group, on what basis are they recommended?

One possibility is that whoever says that we ought to obey certain moral rules is merely expressing his or her subjective preference for obedience to these rules. But then this preference should not, from an impartial point of view, count for any more than any other preferences of similar strength. This desire for obedience to the rules must therefore be weighed against conflicting desires which are frustrated by obedience. Then we are back with the task of impartially adjudicating a conflict of preferences.

If, on the other hand, people say that we ought to obey the moral rules, not because of their subjective preferences, but because these rules are valid independently of any preferences, we must again ask about the nature of the independent moral reality this claim presupposes. The notion of moral rules that are valid independently of their consequences is not a promising basis for challenging conclusions based on an impartial consideration of the interests of all affected, because it unnecessarily imports the puzzling idea of an objective moral reality into an area of human life which, as we have seen, can be explained in a more natural and less mysterious way.

The principle of impartial consideration of interests thus withstands challenges from alternatives which would put ethics on a different basis. It alone remains a rational basis for ethics. This is an important conclusion, so important that the way in which we reached it warrants a brief restatement.

In making ethical decisions I am trying to make decisions which can be defended to others. This requires me to take a perspective from which my own interests count no more, simply because they are my own, than the similar interests of others. Any preference for my own interests must be justified in terms of some broader impartial principle. It might seem that this is compatible with all sorts of moral rules and principles, including some which pay little or no attention to the interests of others, as long as they pay equally little attention to my own interests. When we investigate these other moral rules or principles, however, we find that the grounds for recommending them are either that they will further the interests of all, or simply that they are right in themselves. If the first of these grounds is offered, the principle of equal consideration of interests remains the ultimate basis of morality, and we are left with the task of working out how best to further the interests of all. On the other hand, the idea of moral laws existing independently of the interests and preferences of living beings is implausible, once we have more straightforward explanations of the origins of ethics. Without the notion of an independent moral reality to back them up, however, claims made on behalf of these moral rules or principles can be no more than expressions of personal preferences which, from the collective point of view, should receive no more weight than other preferences. Thus conflicts over differing moral ideals can be treated like any other conflict of preferences, that is, by assessing them impartially and doing what, on the whole, satisfies most preferences.

I have applied this method of argument only to two alternative ethical views—to egoism and to conventional moral rules. But the method is applicable to any value that might be put forward as an alternative to, or a modification of, the principle of equal consideration of interests. It applies to the idea that justice and human rights are ethical principles independent of the equal consideration of interests. Of these principles we can always ask: Is the principle put forward merely as a subjective preference? Then let us give it no more weight than any other preference of equal strength. Is it supposed, on the other hand, to be a true principle, an element of the universe existing independently of us? Then we demand an account of these ethical truths inherent in the universe, how we are aware of them, and why they provide us with reasons for acting, independently of our desires. Until a plausible account has been given—and it has not been given yet—let us cling to the simpler idea that ethics evolved out of our social instincts and our capacity to reason. And let us cling to the principle of equal consideration of interests—which relies on nothing but the fact that we have interests, and the fact that we are rational enough to take a broader point of view from which our own interests are no more important than the interests of others—as a uniquely rational basis for ethical decision-making.

EXPANDING THE CIRCLE OF ETHICS

The idea of impartiality was originally introduced into this discussion because ethics involves justifying one's conduct to one's tribal group or society. The philosophical argument of the last few pages invoked the idea of impartiality but left out the reference to one's own society. This is a significant omission which I must now explain. Obviously there are actions one can defend in a manner that is acceptable within one's own society, but unacceptable to members of other societies. Tribal moralities often take exactly this form. Obligations are limited to members of the tribe; strangers have very limited rights, or no rights at all. Killing a member of the tribe is wrong and will be punished, but killing a member of another tribe whose path you happen to cross is laudable. Nor is this distinction between one's own kind and others limited to illiterate tribes. In the Bible we read the command to the Hebrews:

> When your brother is reduced to poverty and sells himself to you, you shall not use him to work for you as a slave. . . . Such slaves as you have, male or female, shall come from the na-

tions round about you; from them you may buy slaves. You may also buy the children of those who have settled and lodge with you and such of their family as are born in the land. These may become your property, and you may leave them to your sons after you; you may use them as slaves permanently. But your fellow-Israelites you shall not drive with ruthless severity.

Here is a code that could be disinterestedly recommended to Israelites, but hardly to Canaanites. The same general point holds true of ancient Greece, where strangers lacked rights unless they were guests falling under the laws of hospitality. At first this insider/outsider distinction applied even between the citizens of neighboring Greek city-states; thus there is a tombstone of the mid-fifth century B.C. which reads:

> This memorial is set over the body of a very good man. Pythion, from Megara, slew seven men and broke off seven spear points in their bodies . . . This man, who saved three Athenian regiments . . . having brought sorrow to no one among all men who dwell on earth, went down to the underworld felicitated in the eyes of all.

This is quite consistent with the comic way in which Aristophanes treats the starvation of the Greek enemies of the Athenians, starvation which resulted from the devastation the Athenians had themselves inflicted. Plato, however, suggested an advance on this morality: he argued that Greeks should not, in war, enslave other Greeks, lay waste their lands or raze their houses; they should do these things only to non-Greeks.

These examples could be multiplied almost indefinitely. The ancient Assyrian kings boastfully recorded in stone how they had tortured their non-Assyrian enemies and covered the valleys and mountains with their corpses. Romans looked on barbarians as beings who could be captured like animals for use as slaves or made to entertain the crowds by killing each other in the Colosseum. In modern times Europeans have stopped treating each other in this way, but less than two hundred years ago some still regarded Africans as outside the bounds of ethics, and therefore a resource which should be harvested and put to useful work. Similarly Australian aborigines were, to many early settlers from England, a kind of pest, to be hunted and killed whenever they proved troublesome.

So the shift from a point of view that is disinterested between individuals within a group, but not between groups, to a point of view that is fully universal, is a tremendous change—so tremendous, in fact, that it is only just beginning to be accepted on the level of ethical reasoning and is still a long way from acceptance on the level of practice. Nevertheless, it is the direction in which moral thought has been going since ancient times. Is it an accident of history that this should be so, or is it the direction in which our capacity to reason leads us?

Why should our capacity to reason require anything more than disinterestedness within one's own group? Since the interests of my group will often be better served by ignoring the interests of members of other groups, the need for a public justification of conduct should require no more than this. Indeed, shouldn't we rather expect the need for public justification to *prohibit* justifications which give the interests of my group no more weight than the interests of other groups?

This suggestion overlooks the autonomy of reasoning—the feature I have pictured as an escalator. If we do not understand what an escalator is, we might get on it intending to go a few meters, only to find that once we are on, it is difficult to avoid going all the way to the end. Similarly, once reasoning has got started it is hard to tell where it will stop. The idea of a disinterested defense of one's conduct emerges because of the social nature of human beings and the requirements of group living, but in the thought of reasoning beings, it takes on a logic of its own which leads to its extension beyond the bounds of the group.

Here are two historical illustrations of this process. The first is from the laws of inheritance. In many parts of Europe it was once the law that foreigners could not inherit property. If a German who owned property in France died, his property was confiscated. In some countries this even happened to people who had moved merely from one diocese to another. In France this law was abolished in 1790 by the Revolutionary French National Assembly, which considered it contrary to the principle of human brotherhood. This step is a fine example of collective reasoning inspired by the thinkers of the Enlightenment, triumphing over the narrow tendencies of group selection; but, as if to prove that when altruism is not limited to groups, it can still be limited to those who reciprocate, within a few years the Code Napoléon restricted the scope of the abolition to foreigners from those nations which gave reciprocal rights to foreigners within their own territories. Not even reciprocal altruism could hold up progress for long, however, and the restriction was removed in 1819. In England, incidentally, foreigners did not acquire the same rights to inherit property as British subjects until 1870.

My second illustration of the process of reason expanding the bounds of ethics is more recent. It comes from Gunnar

Myrdal's landmark study of the American race question, *An American Dilemma*. The book was published in 1944 and most of the research done before the war, so it describes the situation of American blacks, and the attitudes of whites toward them, long before the successes of the civil rights movement of the sixties. Nevertheless, Myrdal was able to detect a process of ethical reasoning which was putting racist attitudes under pressure. Here are some of the key passages of his description of this process:

> The individual . . . does not act in moral isolation. He is not left alone to manage his rationalizations as he pleases, without interference from outside. His valuations will, instead, be questioned and disputed. Democracy is a "government by discussion," and so, in fact, are other forms of government, though to a lesser degree. Moral discussion goes on in all groups from the intimate family circle to the international conference table. . . .
>
> In this process of moral criticism which men make upon each other, the valuations on the higher and more general planes—referring to *all* human beings and *not* to specific small groups—are regularly invoked by one party or the other, simply because they are held in common among all groups in society, and also because of the supreme prestige they are traditionally awarded. By this democratic process of open discussion there is started a tendency which constantly forces a larger and larger part of the valuation sphere into conscious attention. More is made conscious than any single person or group would on his own initiative find it advantageous to bring forward at the particular moment. . . .
>
> The feeling of need for logical consistency within the hierarchy of moral valuations—and the embarrassed and sometimes distressed feeling that the moral order is shaky—is, in its modern intensity, a rather new phenomenon. With less

mobility, less intellectual communication, and less public discussion, there was in previous generations less exposure of one another's valuation conflicts. The leeway for false beliefs, which makes rationalizations of valuations more perfect for their purpose, was also greater in an age when science was less developed and education less extensive. These historical differentials can be observed today within our own society among the different social layers with varying degrees of education and communication with the larger society, stretching all the way from the tradition-bound, inarticulate, quasi-folk-societies in isolated backward regions to the intellectuals of the cultural centers. When one moves from the former groups to the latter, the sphere of moral valuations becomes less rigid, more ambiguous and also more translucent. At the same time the more general valuations increasingly gain power over the ones bound to traditional peculiarities of regions, classes, or other smaller groups. One of the surest generalizations is that society, in its entirety, is rapidly moving in the direction of the more general valuations.

Myrdal is describing a specific case of the expansion of the moral sphere. Although Myrdal notes ways in which twentieth-century knowledge and communications hasten the process he describes, many of the forces leading to this expansion have operated throughout human history. There may be more moral discussion in a democracy, but there will be some in any community. The feeling of need for consistency among valuations may be more intense in modern times, but has been at least a latent force since humans first became able to recognize inconsistencies. The development of science and the spread of education are relatively recent, but the desire for knowledge on which both are based is not. The only feature of Myrdal's account that has no application to earlier times is his statement that claims referring to *all* human beings are regularly invoked because they are held in common by all groups in society. This has not always been true; but that the wider valuations in the long run attract more support than the narrower is a general tendency, of which Myrdal's case is one example. In Plato's time, to appeal to the claims of "all human beings" would have seemed absurd; but Plato's appeal to consider the welfare of all Greeks, rather than just Athenians, served the same progressive function as the appeal to all humans has served in more recent times.

In studying the history of class revolutions, Marx noted the same expansionist tendency:

> Each new class which displaces the one previously dominant is forced, simply to be able to carry out its aim, to represent its interest as the common interest of all members of society, that is, ideally expressed. It has to give its ideas the form of universality and represent them as the only rational, universally valid ones. . . . Every new class, therefore, achieves dominance only on a broader basis than that of the previous class ruling.

Given his materialist conception of history, Marx would not have admitted that the inherently expansionist nature of reasoning is playing an important *causal* role in the process he describes; rather he thought of it as cloaking the class interests of those making the revolutions. Yet the end result is the same. Marx's own theory leads to the most universal form of human society possible, for he envisaged Communism as a society divided neither by class nor by national boundaries. Though Marx was not impressed by the power of ideas, the idea of universality had a powerful hold on his own thoughts.

We can now state the rational basis of the expansion of ethics. Disinterestedness within a group involves the rejection of purely egoistic reasoning. To reason ethically I have to see my own interests as one among the many interests of those that make up the group, an interest no more important than others. Justifying my actions to the group therefore leads me to take up a perspective from which the fact that I am I and you are you is not important. Within the group, other distinctions are similarly not ethically relevant. That someone is related to *me* rather than to you, or lives in *my* village among the dozen villages that make up our community, is not an ethical justification for special favoritism; it does not allow me to do for my kin or fellow villagers any more than you may do for your kin or fellow villagers. Though ethical systems everywhere recognize special obligations to kin and neighbors, they do so within a framework of impartiality which makes me see my obligations to my kin and neighbors as no more important, from the ethical point of view, than other people's obligations to their own kin and neighbors.

Once I have come to see my interests and those of my kin and neighbors as no more important, from the ethical point of view, than those of others within my society, the next step is to ask why the interests of my society shall be more important than the interests of other societies. If the only answer that can be given is that it is *my* society, then the ethical mode of reasoning will reject it. Otherwise we would simultaneously be holding:

(1) if I claim that what I do is right, while what you do is wrong, I must give some reason other than the fact that my action benefits me (or my kin, or my village) while your action benefits you (or your kin, or your village); and yet

(2)   I can claim that what I do is right, while what you do is wrong, merely on the grounds that my act benefits my society whereas your act benefits your society.

Reasoning beings will not, insofar as they do reason, accept this kind of conflict in their beliefs. If I have seen that from an ethical point of view I am just one person among the many in my society, and my interests are no more important, from the point of view of the whole, than the similar interests of others within my society, I am ready to see that, from a still larger point of view, my society is just one among other societies, and the interests of members of my society are no more important, from that larger perspective, than the similar interests of members of other societies. Ethical reasoning, once begun, pushes against our initially limited ethical horizons, leading us always toward a more universal point of view.

Where does this process end? Taking the impartial element in ethical reasoning to its logical conclusion means, first, accepting that we ought to have equal concern for all human beings. By including fraternity, or the "brotherhood of man," among their ideals along with liberty and equality, the leaders of the French Revolution neatly conveyed the Enlightenment idea of extending to all mankind the concern that we ordinarily feel only for our kin. The ideal of the brotherhood of human beings has now passed into official rhetoric; turning that ideal into reality, however, is another matter. There can be no brotherhood when some nations indulge in previously unheard-of luxuries, while others struggle to stave off famine. To take just one well-known example: because of the extraordinary amount of meat Americans consume—meat which comes from animals fed on grain and soybeans—they are, on average, each responsible for the consumption of 2,000 pounds of grain a year. Indians use only about 450 pounds of grain per year, because they eat their grain directly, instead of cycling it through animals, a process which wastes up to 95 percent of the food value of the grain. What kind of brother would waste so much of his food when his siblings are hungry?

The circle of altruism has broadened from the family and tribe to the nation and race, and we are beginning to recognize that our obligations extend to all human beings. The process should not stop there. In my earlier book, *Animal Liberation*, I showed that it is as arbitrary to restrict the principle of equal consideration of interests to our own species as it would be to restrict it to our own race. The only justifiable stopping place for the expansion of altruism is the point at which all whose welfare can be affected by our actions are included within the circle of altruism. This means that all beings with the capacity to feel pleasure or pain should be included; we can improve their welfare by increasing their pleasures and diminishing their pains.

The expansion of the moral circle should therefore be pushed out until it includes most animals. (I say "most" rather than "all" because there comes a point as we move down the evolutionary scale—oysters, perhaps, or even more rudimentary organisms—when it becomes doubtful if the creature we are dealing with is capable of feeling anything.) From an impartial point of view, the pleasures and pains of non-human animals are no less significant because the animals are not members of the species *Homo sapiens*. This does not mean that a human being and a mouse must always be treated equally, or that their lives are of equal value. Humans have interests—in ideas, in education, in their future plans—

that mice are not capable of having. It is only when we are comparing similar interests—of which the interest in avoiding pain is the most important example—that the principle of equal consideration of interests demands that we give equal weight to the interests of the human and the mouse.

The expansion of the moral circle to non-human animals is only just getting under way. It has still to gain verbal and intellectual acceptance, let alone be generally practiced. Yet the ecology movement has emphasized that we are not the only species on this planet, and should not value everything by its usefulness to human beings; and defenders of rights for animals are gradually replacing the old-fashioned animal welfare organizations which cared a lot for domestic pets but little for animals with less emotional appeal to us. In philosophy departments all over the English-speaking world, the moral status of animals has become a lively topic of debate, and the number of those calling for a change in our present attitude toward animals is growing. The idea of equal consideration for animals strikes many as bizarre, but perhaps no more bizarre than the idea of equal consideration for blacks seemed three hundred years ago. We are witnessing the first stirrings of a momentous new stage in our moral thinking.

Will this new stage also be the final stage in the expansion of ethics? Or will we eventually go beyond animals too, and embrace plants, or perhaps even mountains, rocks, and streams? Since today's enlightened thinking often turns out to be tomorrow's hidebound conservatism—witness the male bias now apparent in the eighteenth-century appeal to "brotherhood"—it would be imprudent to say too firmly that with the inclusion of non-human animals we will at last have gone as far as impartial reasoning requires. Claims that go beyond animal life have been put forward, as part of the general swing away from an ethic concerned only with the welfare of human beings. Albert Schweitzer proposed an ethic of "reverence for life" which specifically included plant life: "A man is really ethical," he wrote, "only when he obeys the constraint laid on him to help all life which he is able to succour, and when he goes out of his way to avoid injuring anything living. He does not ask how far this or that life deserves sympathy as valuable in itself, nor how far it is capable of feeling. To him life itself is sacred." The views of the early ecologist Aldo Leopold are even more relevant here, since Leopold believed that ethics broadens in much the way I have argued it evolves, but saw this expansion as continuing beyond sentient life. Here is a passage from Leopold's *Sand County Almanac:*

> When god-like Odysseus returned from the wars in Troy, he hanged all on one rope a dozen slave-girls of his household whom he suspected of misbehavior during his absence.
>
> This hanging involved no question of propriety. The girls were property. The disposal of property was then, as now, a matter of expediencey, not of right and wrong.
>
> Concepts of right and wrong were not lacking from Odysseus' Greece; witness the fidelity of his wife through the long years before at last his black-prowed galleys clove the wine-dark seas for home. The ethical structure of that day covered wives, but had not yet been extended to human chattels. During the three thousand years which have since elapsed, ethical criteria have been extended to many fields of conduct, with corresponding shrinkages in those judged by expediency only. . . .
>
> There is as yet no ethic dealing with man's relation to land and to the animals and plants which grow upon it. Land, like Odysseus' slave-girls, is still property. The land-relation is still strictly economic, entailing privileges but not obligations.

The extension of ethics to this third element in human environment is, if I read the evidence correctly, an evolutionary possibility and an ecological necessity. It is the third step in a sequence. The first two have already been taken. Individual thinkers since the days of Ezekiel and Isaiah have asserted that the despoliation of land is not only inexpedient but wrong. Society, however, has not yet affirmed their belief.

It is easy to sympathize with Schweitzer's and Leopold's concern for all living things. Once the expansion of ethics to all sentient creatures has been accepted, it is only a small step to extend this expansion until it takes in plants and even inanimate natural objects like the land, streams, and mountains.

Nevertheless, I believe that the boundary of sentience—by which I mean the ability to feel, to suffer from anything or to enjoy anything—is not a morally arbitrary boundary in the way that the boundaries of race or species are arbitrary. There is a genuine difficulty in understanding how chopping down a tree can matter *to the tree* if the tree can feel nothing. The same is true of quarrying a mountain. Certainly imagining myself in the position of the tree or mountain will not help me to see why their destruction is wrong; for such imagining yields a perfect blank. Often it will be wrong to reduce a mountain to gravel because of the loss of aesthetic or recreational values, or because it deprives thousands of animals of their habitat; but can it be wrong *in itself*, apart from all these effects on creatures capable of suffering and enjoyment?

Perhaps my incomprehension proves only that I, like earlier humans, am unable to break through the limited vision of my own time. Yet there is a sense in which the limits of sentience are not really limits at all, for applying the test of imagining ourselves in the position of those affected by our actions shows that in the case of nonsentient things there is nothing at all to be taken into account. We need not deliberately exclude nonsentient things from the scope of the principle of equal consideration of interests: it is just that including them within the scope of this principle leads to results identical with excluding them, since they have no preferences—and therefore no interests, strictly speaking—to be considered. There is nothing we can do that matters to them. Whatever consideration we give to something with no interests still leaves us with nothing. That is why I believe that if ethics grows to take into account the interests of all sentient creatures, the expansion of our moral horizons will at last have completed its long and erratic course.

5

REASON AND GENES

Man is a reasoning animal.

—SENECA, *Ad Lucilium*

Man is a rational animal who always loses his temper when he is called upon to act in accordance with the dictates of reason.

—OSCAR WILDE, *The Critic as Artist*

The previous chapter concluded on a lofty note. Now we must descend to earth. We are capable of reasoning; but we are also the products of selective pressure on genes. We owe our existence to the ability of our ancestors to further their own interests and the interests of their kin. Can we really expect beings who have evolved in this manner to give up their narrower pursuits and adopt the universal standpoint of pure reason?

In 1739 David Hume's *Treatise of Human Nature* challenged the common view that our reason and our desires are locked in combat. Hume declared that reason cannot conflict with desires, because reason can do no more than give direction to desires that already exist; it is powerless against our existing desires and cannot give rise to new desires. What we may think of as a conflict between reason and passion—

like the choice between a romantic but worthless lover and a less exciting but more deeply fulfilling marriage—is really a conflict between intense, but short-term, and calmer, but more enduring, desires. Reason may help us to sort out the consequences of our choice, but it cannot tell us what we most want. "Reason is, and ought only to be," wrote Hume, "the slave of the passions."

Hume's statement was a deduction from his view of the nature of reason, rather than a speculation drawn from observation or science; yet it is recognizably the ancestor of the view some sociobiologists take of human behavior. "Reason is, and ought only to be," they would rewrite Hume, "the slave of our genes." On this basis they would dismiss as philosophical fantasy the idea that reason draws us toward a universal point of view.

Nor are sociobiologists the only ones who will be skeptical of the conclusions of the previous chapter. In an earlier chapter I noted that virtually no one gives as much consideration to the interests of strangers as they give to their own interests or those of their families. Yet most of us are capable of reasoning. In deciding how best to further our own interests and those of our families, we do quite a lot of reasoning. Doesn't this show that reason operates to serve our self-interested desires, not to lead us out to wider concerns? Isn't reason incapable of altering our fundamentally selfish concerns?

SELFISHNESS

As long as there have been philosophers, and no doubt before, some have thought that everything anyone does is ultimately selfish. Philosophers call this view "psychological egoism" because it asserts, as a matter of psychological fact, that people behave egoistically. Nearly all philosophers now reject the doctrine. They point out that those who hold it must choose between one of two interpretations of "selfish." In the first interpretation, to be selfish is to take no account of the interests of anyone else, except when by doing so you can get more of what you want for yourself. This is roughly what we usually mean when we say that someone is selfish, but it is very implausible to say that *everyone* is *always* selfish in this sense. There are examples of people who do things for others with no prospect of reward, ranging from patriots who die for their country to volunteers who donate a pint of their blood to help a stranger. So psychological egoists often take a broader view of "selfish" behavior. They say that if patriots volunteer for suicidal missions, that must show that they want to die for their country more than they want to go on living; and if blood donors give blood at no fee to a stranger, that must be because they get satisfaction from helping strangers. In this second interpretation of "selfish" it is much more difficult to refute the claim that everyone always acts selfishly—but now that claim has changed its meaning so radically that it is no longer the bold challenge to more idealistic theories of human nature that it at first seemed to be. This redefined version of psychological egoism is quite compatible with distinguishing between behavior that is selfish in the ordinary sense of the term and behavior that is "selfish" only in the peculiar sense in which the person who would rather help others than see them suffer is selfish. This second sense of "selfish" is so all-encompassing that it serves no useful function at all. If *all* behavior is, in this sense, selfish, but some is also selfish in the first, narrower sense, clarity will be best served by restricting the term to its narrower meaning, which has the advantage of contrasting some kinds of behavior with others.

Though this refutation of psychological egoism has often been made, and is, I believe, entirely sound, the doctrine refuses to die. It has resurfaced in the writings of sociobiologists, which is ironic because sociobiology, properly understood, provides a clear reason for rejecting psychological egoism.

The evolutionary theories of sociobiologists show that beings who considered only their own interests would leave fewer descendants than beings who also considered the interests of their kin. So there is a good reason to believe that we do not all act solely in our own interests. Genes promoting strictly selfish behavior in individual animals would be less likely to survive than genes which do not.

A sociobiological approach can be taken still further in opposition to psychological egoism. As we saw in Chapter 2, the existence of real-life Prisoner's Dilemma situations puts egoists at a disadvantage in situations where cooperation is advantageous. In these situations two genuine altruists will do better than two egoists, and a single egoist will not do as well as an altruist if her egoism is apparent to others. So at least within the sphere of personal relationships, genuine altruism could have come about consistent with the theory of evolution.

Despite this, sociobiologists often sound like psychological egoists. In *The Selfish Gene*, Richard Dawkins writes: "If you look at the way natural selection works, it seems to follow that anything that has evolved by natural selection should be selfish." Edward Wilson says: "When altruism is conceived as the mechanism by which DNA multiplies itself through a network of relatives, spirituality becomes just one more Darwinian enabling device." If this is selfishness, though, it is a strange form of it. For sociobiologists, a being acts altruistically if it increases the fitness of another at the expense of its own, and selfishly if it increases its own fitness at the expense of another's. This looks like normal usage, until we realize that by "fitness" sociobiologists do not mean, as we might at first imagine, the individual's own fitness for survival; instead they mean fitness *as measured by the number of surviving offspring*. Here is the peculiarity of the sociobiological idea of selfishness. Mary Midgley remarks in *Beast and Man* that if a person is described as thinking of nothing but how many descendants he will have in five centuries, we would be more inclined to call him crazy than selfish. Midgley is right, but the sociobiologists' sense of "selfish" is even more peculiar than that, for it implies that a person is selfish if he acts in a way that in fact will maximize the number of descendants he will have in five centuries, *although he is all the while thinking only of the welfare of others!* In the writings of sociobiologists, "selfish" and "altruistic" have nothing to do with motivation; they refer only to the actual consequences of the individual's behavior, whether or not the individual is motivated by or even aware of these consequences. That is why Dawkins can write of a "selfish gene" and Wilson of an altruistic parasite. Using these terms in this way makes genetics and the study of parasites more readily understandable, but to transfer this usage to discussions of human behavior without noting that "selfish" genes are entirely compatible with completely unselfish motivation on the part of those whose genes they are, would be highly misleading.

So sociobiology, properly understood, does not support the view that we are all irredeemably selfish, at least not in any normal sense of the term. Does it, though, suggest that we are bound to limit our altruism to our kin, or to those who will help us when we need it, or at most to those in some narrow group to which we belong? Wilson seems to think so: he accounts for Mother Teresa's work caring for the poor and ill of Calcutta by referring to her belief in Christianity and its doctrine of reward in the afterlife. Garrett Hardin, also arguing on biological grounds, writes that altruism can only exist "on a small scale, over the short term, in certain circumstances and within small, intimate groups." Dawkins says: "Much as we might wish to believe otherwise, universal love and the welfare of the species as a whole are concepts which simply do not make evolutionary sense."

If these statements are intended to deny the possibility of universal altruism, they go beyond what sociobiological theories justify. (As we shall see, Dawkins, at least, may not intend to deny this possibility.) As individuals, we do many things that make no evolutionary sense. Take sex as an example. The strength and prevalence among human beings of the desire for sexual intercourse is doubtless due to evolution. Humans with little sexual desire had less chance of passing on their genes. The central evolutionary function of this desire is reproduction. That humans have sex even when females are not fertile suggests that sex has other evolutionary functions as well, probably connected with the desirability of lasting relationships to provide for the offspring during their long period of dependence, but it does not count against reproduction being the central evolutionary function of sexual desire. Yet millions of human beings take great care that their sexual activities will not lead to reproduction. This is not, by and large, because they are pursuing the evolutionary strategy of putting a lot of care into a small number of children to ensure that they survive and reproduce. Wealthy people who could easily rear ten children rarely do so. Some choose to have no children at all.

The use of contraceptives by people who could successfully rear more children to adulthood may not make "evolutionary sense," but it is a practice that shows no sign of becoming extinct. (Don't say that it makes evolutionary sense because the population explosion makes it necessary for the survival of the species—that invokes the mistaken idea of selection on the level of species, not genes.) The lesson to be drawn from the spread of contraception is that reasoning beings are not bound to do what makes evolutionary sense—a point that sociobiologists generally admit, only to give it little weight in their further speculations. The growth of modern contraceptive techniques is a splendid example of the use of reason to overcome the normal consequences of our evolved behavior. It shows that reason can master our genes.

Two objections to this conclusion need to be considered. The first is that if the use of contraception is contrary to evolutionary principles, it will in the long run eliminate itself. People who do not use contraceptives will gradually come to outnumber, and eventually replace entirely, people who do use contraceptives. This dismal prophecy is based on the implausible assumption that individuals' decisions to use or not to use contraception are determined largely by their genes. It seems much more likely that upbringing and education play the major role. There are reasons why people use contraceptives, and in the appropriate circumstances anyone capable of a minimal level of thought may come to see contraception as a good idea. Since there are good evolutionary reasons why a minimal capacity for reasoning is unlikely to be bred out of the human species, the use of contraception is also unlikely to eliminate itself.

The second objection is more profound. It is that the reasoning employed in developing and using contraceptives does not master our genes, because it is all directed at allowing us to fulfill our genetically based desire for sexual intercourse. Admittedly, the objection continues, in the case of sex we can divide the satisfaction of the desire from the original evolutionary function of the desire, but that is only because, in the case of sex, evolution works indirectly. It implants in us a desire for sex that is distinct from a desire for having children. We have learned how to have one without the other. But this has no bearing on the ability of reason alone to overcome our genes; and in particular it does not show that reason can overcome the genetic tendency to limit our altruistic behavior to our kin or some other narrow circle, for this is a case in which evolution works more directly, causing us to have desires for the welfare of our family and compatriots that are much stronger than our desires for the welfare of strangers.

This second objection indicates the need for caution. We should not take our success in counteracting evolution's way of promoting reproduction as an indication that we can as easily overcome evolution's limits on altruism. Maybe evolved impulses can only be overcome when we have other evolved impulses to build on; maybe not. In explaining the importance of understanding our biology, Dawkins writes; "Let us understand what our own selfish genes are up to, because we may then at least have the chance to upset their designs, something which no other species has ever aspired to." This is reminiscent of T. H. Huxley's view that ethical progress depends on combating the process of evolution. To what extent this can be done we do not really know, although we know more about it than Huxley did. What can be said about human selfishness is that there is no reason to believe that we always do what is in our own interest, whether we take this term either in the usual sense of getting more of what we want for ourselves or in the extended biological sense of enhancing the survival of our genes. We can therefore go on to consider with an open mind the possibility of rationally based altruism.

UNSELFISHNESS

In Britain, blood needed for medical purposes comes exclusively from people who voluntarily give their blood to the National Blood Transfusion Service. These donors are not paid. They do not get preferential treatment when they themselves need blood, for the National Health Service provides blood free of charge for all those in Britain who need it. Nor can donors be rewarded—or even given a grateful smile—by the patients whose lives are saved by their gifts. Donors never know who receives their blood, and patients never know who gave the blood they receive.

Common sense tells us that people who give blood do it to help others, not for any disguised benefit to themselves. Richard Titmuss asked thousands of donors why they first gave blood, and the answer supported this common-sense view: less than 2 percent gave explanations that smacked of self-interest, for instance, the belief that giving blood makes you healthier. Allowing for a natural tendency to overstate the extent to which one is altruistically motivated, the conclusion still seems inescapable that this system of obtaining blood depends on altruistic donors. Nor can these gifts be dismissed as isolated, freak occurrences. The National Blood Transfusion Service meets the medical needs of more than 60 million people. It has existed for thirty years, and draws on over a million donors each year.

The British National Blood Transfusion Service is not unique; similar systems exist in Australia, Holland, and some other countries. These systems are working refutations of the contention that altruism can only exist among kin, within small groups, or where it pays off by encouraging reciprocal altruism. There are other instances of equally altruistic behavior, though the lack of contact between donor and recipient makes this one unusually clear-cut. Anyway, one widespread practice involving people helping others without hope of reward is enough. Genuine, non-reciprocal altruism directed toward strangers does occur.

What are the implications of the existence of altruism for the theories of human nature we have been considering? Any theory which entails that non-reciprocal altruism toward strangers cannot occur must be wrong. Does this mean that the evolutionary theories of the origin of altruism discussed in the first chapter of this book must be wrong? It may seem that it does, since these theories explained the rise of kin altruism, reciprocal altruism, and possibly a little group altruism, but could not account for altruism to strangers who cannot reciprocate. But recall the argument of the preceding chapter in which I suggested that altruistic impulses once limited to one's kin and one's own group might be extended to a wider circle by reasoning creatures who can see that they and their kin are one group among others, and from an impartial point of view no more important than others. Biological theories of the evolution of altruism through kin selection, reciprocity, and group selection can be made compatible with the existence of non-reciprocal altruism toward strangers if they can accept this kind of extension of the circle of altruism.

To be sure, this explanation of the broadening of the circle of altruism is not the only possible account. Earlier in this century Edward Westermarck noted the tendency of the circle of morality to expand, but he attributed it not to our capacity to reason, but to an expansion of the altruistic sentiments that he thought were the foundation of all morality. He pointed to the increasing size of our community—from the village to the nation, and now to the world as a whole—as a factor in the breakdown of narrower limits to our concerns and sympathies.

Should we accept the account I have offered rather than an account like Westermarck's? We do not have to choose one or the other; we can accept both explanations. The expansion of the community must have played a role in the expansion of altruism. Once one group starts to interact with another, perhaps hunting or gathering food together, or exchanging goods, the advantages of a reciprocal altruism begin to play a role between groups as well as within each group. So notions of gratitude, of fairness, and of not harming those who do not harm you may extend beyond the group. The plausibility of this account of the expansion of the circle of morality, however, is no ground for denying a role to reason. For it is independently plausible that reasoning should lead us to a more and more universal view of ethics. It is plausible—as I have already argued—in view of the nature of reason and the way in which it logically extends itself beyond narrow bounds. It is also plausible in the light of what we know about the development of ethical thought in a wide variety of cultures.

The idea of an impartial standard for ethics has been expressed by the leading thinkers of the major ethical and religious traditions. In Judaism the rule is to love your neighbor as yourself; a rule which Jesus elevated to the status of one of the two great commandments. About the same time, Rabbi Hillel said: "What is hateful to you do not do to your neighbor; that is the whole Torah, the rest is commentary thereof." Jesus also put it another way: "As you would that men should do to you, do ye also to them likewise." When Confucius was asked for a single word which could serve as a rule of practice for all one's life, he replied: "Is not reciprocity such a word? What you do not want done to yourself, do not do to others." In Indian thought we find the Mahabharata saying:

> Let no man do to another that which would be repugnant to himself; this is the sum of righteousness; the rest is according to inclination. In refusing, in bestowing, in regard to pleasure and to pain, to what is agreeable and disagreeable, a man obtains the proper rule by regarding the case as like his own.

Among the Stoic philosophers of the Roman Empire, Marcus Aurelius argued that our common reason makes us all fellow citizens, and Seneca claimed that the wise man will esteem the community of all rational beings far above any particular community in which the accident of birth has placed him.

It hardly seems necessary to follow the progress of this idea into modern times, where it has become central to popular moral teaching as well as to the ethical writings of a wide range of contemporary philosophers. That the idea of treating others as one would like to be treated oneself should often be repeated is not surprising; what is surprising is the way in which the idea crops up independently in quite different ethical and cultural traditions and is, in each case, seized on as something fundamental to ethical living, a foundation from which all else can be derived. Or rather: this would be surprising if reason had no rule to play in ethics. If ethics were simply the product of our evolved tendencies to help our kin, those who help us, and perhaps our own small group, the fact that ethical teachers have, again and again, independently emphasized a higher and wider standard of conduct would be puzzling. (Note that the reciprocity these ethical thinkers advocate is not that which the "reciprocal altruism" of the sociobiologists would lead us to expect—it is not a recommendation that we do to others as they have done to us, but that we do to them what we would wish them to do to us. Nor is anything said about doing this only if they are likely to respond in kind.) Once reason is admitted to have a role to play in ethics, however, there is nothing at all surprising in the fact that, despite immense cultural differences, outstanding thinkers in different periods and places should extrapolate beyond more limited forms of altruism to what is essentially the same fundamental principle of an impartial ethic.

To the historical and cross-cultural evidence for an association between reasoning and the expansion of the circle of morality, we can add a theory which suggests that the same process occurs on an individual level, as children mature. Lawrence Kohlberg, following up the psychological theories of development postulated by Jean Piaget, asserts that children go through a definite sequence of stages of moral thinking as they develop. In broad terms, the sequence moves from an egoistic level at which morality is seen as a matter of reward and punishment, through a second level of loyalty to group standards, to a third level which seeks out a basis for moral principles which is independent of either self-interest or group standards. Kohlberg believes that there is a logical order in this progression, each level having a higher logical structure than the one before it. Thus certain reasoning skills—for instance, the ability to imagine oneself in the position of another—are necessary for moving to a higher level. Kohlberg does not claim that the ability to reason well is a sufficient condition for being on the highest moral level, but that it is a necessary condition. He also claims that there is a demonstrable connection between moral reasoning and moral action. All these claims accord with the case for regarding reason as the key factor in the spread of altruism beyond its biologically predictable limits.

Independent studies partially support Kohlberg's theories. It has been shown that older children are more likely to be generous than younger ones, and within a given age group, children who do better at tests of the capacity to take on the role of another are more generous than children who do not do so well at these tests. There is data favoring Kohlberg's claim that being able to reason well is a necessary, but not a sufficient, condition for being at one of the higher moral stages at which greater attention is given to the interests of others. There is also evidence linking moral reasoning to moral behavior—for instance, delinquents rank lower on Kohlberg's scale of moral reasoning than other children, while children who score well on a test of moral judgment devised by Piaget have been found more likely to share candy or help other children with a difficult task.

The evidence is not, however, clear-cut. A few studies find no correlation between reasoning ability and moral level, or between moral level and altruistic actions. Moreover, Kohlberg's own data is open to serious objection: for instance, his system of scoring people at higher or lower stages of moral development may already have within it a bias toward higher scores for those with better verbal and reasoning abilities.

Kohlberg's theory is in line with the overall view of the role of reason in ethics I am defending, but at this stage it is an interesting speculation rather than an established fact.

Another argument for the view that reasoning has helped to push out the boundary of altruism is that on this hypothesis it is possible to give a convincing answer to the question: Why hasn't evolution eliminated genuine non-reciprocal altruism toward strangers? Whenever altruism to strangers appeared, why didn't those who had this idea go under in the struggle for survival, taking their extended altruism with them?

This is the familiar puzzle of altruism which sociobiologists have set out to solve; but they have solved it only for altruism toward kin or toward those who can reciprocate. If we say, as Westermarck did, that the expansion of the sphere of altruism has been solely the result of an expansion of the *feeling* of benevolence, the existence of genuine, non-reciprocal altruism toward strangers remains mysterious. Evolution should have wiped out such non-rewarding traits as a broad, unselfish feeling of benevolence. If, however, we say that the expansion of the sphere of altruism is the result of the human capacity to reason, a possible solution to the mystery emerges. For the capacity to reason is not something that evolution is likely to eliminate. In finding food, in avoiding danger, in every area of life, those who reason well have an immense advantage over similar beings less capable of reasoning. So we can expect evolution to select strongly for a high level of reasoning ability. (We know that the human brain did grow with remarkable speed.) Accordingly, if the capacity for reasoning brings with it an appreciation of the reasons for extending to strangers the concern we feel for our kin and our friends, evolution would not eliminate this rational appreciation of the basis of ethics. The price would be too high. The evolutionary advantages of the capacity to reason would outweigh the disadvantages of occasional actions which benefit strangers at some cost to oneself. Hence while the persistence of genuine altruism would be inexplicable if it were based on feeling alone, it becomes much easier to understand if it is not feeling, but reason that is chiefly responsible for it.

Of course, this assumes that coming to see the validity of the reasons for extending concern beyond the narrow circle of friends and relations is closely linked with the capacity to reason at a high level. If the link could easily be severed, evolutionary pressure would sever it. The picture of reasoning developed in Chapter 4 suggests, however, that the ability to reason and the ability to see the reasons for a wider moral concern are essentially the same ability. Just as any person who can reason adequately can, like Hobbes, follow Euclid's proofs of the theorems of geometry, so can anyone capable of reasoning understand the objective point of view from which his or her interests are no more important than the like interests of anyone else. As Humphrey Bogart puts it in that great modern morality tale *Casablanca:* "Look, I'm no good at being noble, but it doesn't take much to see that the problems of three little people don't amount to a hill of beans in this crazy world."

AMBIVALENCE

It may not take much to *see* that one's own problems are the problems of just one among many, but seeing this is not the same as acting in accordance with it. I can accept that from an objective viewpoint my interests count for no more than yours, and at the same time I can disregard this objective standpoint and give my own interests the priority they have from my own subjective standpoint. In *Casablanca*, Bogart lets the woman he loves go, showing that under his cynical exterior he really is good at being noble; but no movie-goer would have been surprised if Bogart had found the insights of the objective standpoint more resistible than the charms of Ingrid Bergman. Many people in no way deficient in their ability to reason rarely or never act in accordance with the objective standpoint. They usually or always give priority to their own interests. To revert to the example of altruism given earlier, while many people in Britain do give blood to strangers, far more—94 percent, to be precise—do not. Undoubtedly many of those who do not give blood reason as well as or better than some of those who do. Notoriously, among the ranks of cheats, swindlers, and other criminals we can find people who reason brilliantly to achieve their selfish goals. These people are capable of following the line of reasoning that leds to altruism, yet they do not do so, or if they do, they disregard it in their actions.

If people are capable of grasping the reasons for taking an objective point of view, how do we explain the fact that many act as if these reasons did not exist? Here we return to Hume's thesis, with which this chapter began, that reason is the slave of the passions. If Hume were right, it would not be surprising that few people act altruistically toward strangers. Only those who happened to have altruistic desires for the welfare of strangers would act in this manner, and for now familiar evolutionary reasons, we should expect few people to have this desire.

Hume was at least partly right. Alone and unaided, reason

cannot give rise to action. There must be some desire, some want or aversion, some pro or con feeling with which reason can combine to generate an action. Hume was right to regard reason as a tool for obtaining what one wants. When people single-mindedly desire to further their own interests, or those of their families, reason serves only to help them get what they want.

On the other hand, Hume was not entirely right. Tools have a way of influencing the purpose for which they are used, especially if that purpose is pursued with less than single-minded determination. The automobile was developed as a means of transporting oneself from A to B. Now people spend their Sunday afternoons in a form of recreation known as "going for a drive"; they travel for an hour or two and then return home without having left the car. The purpose of the outing was to drive, not to go somewhere. Something similar can happen with the tool of reasoning.

We built cars to travel and then discovered that we enjoyed driving for its own sake. In the case of ethical reasoning, we begin to reason impartially in order to justify our conduct to others, and then discover that we prefer to act in accordance with the conclusions of impartial reasoning. Recall Gunnar Myrdal's analysis of white American attitudes about racial inequality; what Myrdal describes as "the feeling of need for logical consistency within the hierarchy of moral evaluations" may be a *feeling,* but it is a feeling that derives from our capacity to reason—of which our capacity to recognize inconsistency is a part. The recognition of inconsistency comes first, and the feeling that this should be avoided follows from it.

That human beings are uncomfortable with inconsistencies is easily observable, and explicable by reference to the decisive role our capacity to behave rationally has played in our evolution. Psychologists use the impressive label "cognitive dissonance" for this phenomenon. Leon Festinger, in *A Theory of Cognitive Dissonance*, sums it up this way:

> In short, I am proposing that dissonance, that is, the existence of nonfitting relations among cognitions, is a motivating factor in its own right. By the term *cognition* ... I mean any knowledge, opinion, or belief about the environment, about oneself, or about one's behavior. Cognitive dissonance can be seen as an antecedent condition which leads to activity oriented toward dissonance reduction just as hunger leads to activity oriented toward hunger reduction.

Which, being translated, means that if we sense an inconsistency in our beliefs, or between our beliefs and our actions, we will try to do something to eliminate the sense of inconsistency, just as when we feel hungry we will try to do something to eliminate our hunger. There are, as Myrdal noted and Festinger also points out, several ways to eliminate a sense of inconsistency, of which making our beliefs and actions both true and consistent is only one. We can, for instance, accept an otherwise implausible belief which reconciles our inconsistent beliefs; or we can drop all interest in the area in which the inconsistency occurs. Human beings are not the perfectly rational creatures they would be if they strove for truth and consistency at all times. Nevertheless, if we can be motivated by a desire to eliminate inconsistency in our beliefs and actions, reason is no mere slave. We may use reason to enable us to satisfy our needs, but reason then develops its own motivating force.

Whether particular people with the capacity to take an objective point of view actually do take this objective viewpoint into account when they act will depend on the strength of their desire to avoid inconsistency between the way they reason publicly and the way they act. The strength of a desire is a relative matter, dependent on the strength of desires which pull in different directions. Our desires to further our own interests and those of our immediate families are biologically older than our desire to avoid inconsistency, for our ancestors had to survive, and help their offspring to survive, long before they had evolved to the point at which they could have inconsistent beliefs. The older desires can conflict with the newer ones, and often the older ones will prove stronger. There is a sense of the word "rational" in which we are rational if we act so as to achieve what, on balance, we desire most. In this sense of the term, people can be perfectly rational and yet perfectly self-interested.

There is a further reason why the desire to avoid inconsistency may not be sufficient to counteract self-centered desires. We are dealing not with a straightforward inconsistency between an individual's beliefs and his or her actions, but with a clash between the individual's actions and the principles he or she must publicly espouse. One can adopt one set of principles in private and a different set in public without any inconsistency; all one has to do is make one's overriding principle the pursuit of self-interest, and then use ethical reasoning in public situations for the purpose of impressing others with one's impartiality, but not as a real guide to one's actions. This is hypocritical, but the hypocrisy is part of a consistent design for promoting one's own interests.

Nevertheless, because we are social beings, reared and educated in a community and bound to the community by deep emotional ties, a life of systematic hypocrisy is likely to be uncomfortable. To present a false face in public, to be constantly on guard instead of open and spontaneous, to deceive even one's friends about one's true principles—all this brings disharmony into one's life. The desire to reduce this disharmony between public principles and private practice could operate as a motivating force, much like the desire to eliminate a straightforward inconsistency between beliefs and actions. While we might, in theory, eliminate the disharmony by taking a totally self-centered stance and regarding the public standpoint with cynical disdain, in practice most of us have too much natural sympathy for others, and too many emotional ties with our community, to take this course.

There is another reason for not choosing a totally self-centered life. Since ancient times, philosophers have maintained that to strive too hard for one's own happiness is self-defeating. The "paradox of hedonism," as philosophers have called it, is that those who seek their own pleasure do not find it, and those who do not seek it find it anyway. The pleasures of a self-centered life eventually pall and the drive for still higher levels of luxury and delight brings no lasting satisfaction. Real fulfillment is more likely to be found in working for some other end. Hence, these philosophers claim, if we want to lead a happy life, we should not seek happiness directly, but should find a larger purpose in life, outside ourselves.

The claim that we are more likely to find happiness if we pursue it indirectly is, of course, a psychological generalization and like most such generalizations it does not hold for everyone; but it is true that the lives of those who have nothing to do but enjoy themselves are much less happy than we would expect them to be if human nature were suited to the unalloyed pursuit of personal pleasure. Perhaps, having developed into beings with purposes, we are naturally driven to seek larger purposes, which give meaning and significance to our lives. Perhaps the boredom and loss of interest in life observable in many of those with no purposes beyond their own pleasure are the result of neglecting this aspect of our nature. If this speculation is correct, the logical possibility of consistently rejecting the claims of the ethical point of view becomes much less attractive as a practical option. The impartial standpoint of ethics has the advantage of drawing us beyond a concern with our own interests, to wider purposes in which we can find deeper fulfillment and a more meaningful life.

For these reasons, few of us will dismiss the ethical point of view altogether. Yet our self-centered desires (including our desires for our kin and close friends) remain strong. The result is a tension between these desires and our ethical commitment. Edward Wilson has said that the theory of group selection "predicts ambivalence as a way of life in social creatures." He was thinking of the conflicting loyalties between different levels of selection, particularly self, family, and tribe. In social creatures who reason, this ambivalence can take the form of a conflict between our self-centered desires and our desire to act in accordance with the standards of public justification that we invoke as a member of a group. While on the collective level the ability to reason brings with it the rational basis of an expanding altruism, on the individual level it need not do so. Wilson's remark is a biologist's recapitulation of a division in human nature that philosophers have often noticed. The difference is that whereas philosophers like Plato and Kant have seen the conflict as one between our reason and our desires, Wilson is closer to Hume's view that it is a conflict between self-interested desires and desires like sympathy and benevolence, with reason standing on the sidelines powerless to intervene. I have suggested that reason is not powerless. On the collective level, once we have begun to justify our conduct publicly, reason leads us to develop and expand our moral concerns, drawing us on toward an objective point of view. On the individual level reason is less compelling; while it leads us to see inconsistencies between our beliefs and our actions, or between what we profess in public and what we do in private, the desire to avoid these inconsistencies is not always strong enough to overcome other desires. As a result, reason can get channeled into narrower pursuits than we can justify from an objective standpoint. The shape of human ethical systems is an outcome of the attempt of human societies to cope with this tension between collective reasoning and the biologically based desires of individual human beings.

6

A NEW UNDERSTANDING OF ETHICS

... the soundest criterion of virtue is, to put ourselves in the place of an impartial spectator, of an angelic nature, suppose, beholding us from an elevated station and uninfluenced by our prejudices, conceiving what would be his estimate of the intrinsic circumstances of our neighbour, and acting accordingly.

—WILLIAM GODWIN,
Enquiry Concerning Political Justice

We are afraid to put men to live and trade each on his own private stock of reason; because we suspect that this stock in each man is small, and that the individuals would do better to avail themselves of the general bank and capital of nations, and of ages. Many of our men of speculation, instead of exploding general prejudices, employ their sagacity to discover the latent wisdom which prevails in them. If they find what they seek, and they seldom fail, they think it more wise to continue the prejudice, with the reason involved, than to cast away the cost of prejudice, and to leave nothing but the naked reason; because prejudice, with its reason, has a motive to give action to that reason, and an affection which will give it permanence. Prejudice is of ready application in the emergency; it previously engages the mind in a steady course of wisdom and virtue, and does not leave the man hesitating in the moment of decision, sceptical, puzzled, and unresolved. Prejudice renders a man's virtue his habit; and not a series of unconnected acts. Through just prejudice, his duty becomes a part of his nature.

—EDMUND BURKE,
Reflections on the Revolution in France

SCIENCE AND MORAL INTUITIONS

The account of ethics sociobiologists offer is incomplete and therefore misleading. Nevertheless, sociobiology provides the basis for a new understanding of ethics. It enables us to see ethics as a mode of human reasoning which develops in a group context, building on more limited, biologically based forms of altruism.

So ethics loses its air of mystery. Its principles are not laws written up in heaven. Nor are they absolute truths about the universe, known by intuition. The principles of ethics come from our own nature as social, reasoning beings. At the same time, a view of ethics grounded on evolutionary theory need not reduce ethics simply to a matter of subjective feelings or arbitrary choices. The fact that our ethical judgments are not dictated to us by an external authority does not mean that any ethical judgment is as good as any other. Ethical reasoning points the way to an assessment of ethical judgments from an objective point of view.

Discussions of ethical questions today are often confused and irresolute because those taking part are muddled about the foundations of ethics. In principle, we might expect that our new understanding of ethics should clear up these muddles and make agreement easier to reach in ethics. But if biology cannot furnish us with ethical premises, how will this happen?

The gap between facts and values exists because while we do not choose the way the world is, we do choose what we are going to do. If this choice were totally subjective, the gap between facts and values would open very wide. Emphasizing the rational element in ethical choice, however, narrows the gap between facts and values. Rational criticism of an ethical choice becomes possible, and facts may be relevant to this rational process.

This does not mean that we can, after all, derive ethical principles from biology. Discovering that some form of behavior has a biological basis does not justify that kind of behavior. Sometimes, as we saw in Chapter 3, the effect of this discovery will be the very opposite of a justification. Learning that what we have taken to be a self-evident moral rule has a biological explanation should lead us to question our acceptance of the moral rule.

At the end of Chapter 3, I suggested that this debunking effect of biological and cultural explanations of our ethical principles would not be possible if *all* our ethical beliefs could be accounted for in these ways. Then all of our ethical principles would be on an equal footing, and since we still have to decide what to do, we could not take a skeptical attitude toward an ethical principle merely because we knew of a biological explanation for our holding it. Only if there is a rational component to ethics, I argued, can we use biological explanations to distinguish the rational elements of our moral principles from the biological elements.

If the argument of Chapter 4 is sound, there is a rational component to ethics. Taking an objective point of view involves seeing our own interests as no more important than the like interests of anyone else. This yields the principle of equal consideration of the interests of all. If this, and this alone, is the rational component of ethics, there should be a debunking explanation—biological or cultural—for every other aspect of our conventional ethical beliefs, from trite moral rules against lying and stealing to such noble constructions as justice and human rights. If so, when the debunked principles have been scrutinized, found wanting, and cleared away, we will be left with nothing but the impartial rationality of the principle of equal consideration of interests.

Could so drastic a demolition of conventional morality be right? Or would it make ethics so abstract and divorced from human nature as to be—to use Wilson's description of Rawls—"an ideal state for disembodied spirits" but totally inapplicable to real human beings?

Very few philosophers have been prepared to take impartiality all the way to its logical conclusion. One of the few was William Godwin, the eighteenth-century anarchist author of the *Enquiry Concerning Political Justice* and, incidentally, the husband of the feminist Mary Wollstonecraft and the father of Mary Shelley, the author of *Frankenstein*. In an example which has been argued about ever since it first appeared in his *Enquiry*, Godwin asks you to imagine that Fénelon, archbishop of Cambrai and then a famous author, is trapped in a burning building along with his valet, who happens to be your father. There is time to save only one of them. Who is it to be? Godwin says you should rescue Fénelon, for the books he writes bring wisdom and joy to thousands. His life is thus more valuable than the life of the valet; and that the valet is your father is irrelevant. As Godwin puts it: "What magic is there in the pronoun 'my' that should justify us in overturning the decisions of impartial truth?" Instead Godwin proposes as "the soundest criterion of virtue" the judgments that would be made by an impartial spectator not influenced by our "prejudices."

Godwin's criterion of virtue stands on the firm basis of impartial reason, the same standard of impartiality defended in the preceding chapters of this book. Nevertheless, describing affectionate feelings for one's father as "prejudice" seems an excessively abstract form of reasoning, divorced from the concrete realities of human life.

Godwin himself came to appreciate that "prejudice" was the wrong word. A few years after the publication of the *Enquiry* its doctrines were virulently denounced by the cleric Samuel Parr. In a reply to Parr, Godwin revised his position about Fénelon and the valet. He noted that we would scarcely blame someone who saved his father rather than the archbishop, because "the sentiment of filial affection . . . is a feeling pregnant with a thousand good and commendable actions" and is a sign of a virtuous and honorable nature. In saying this Godwin was not retracting his view that it would be better to save Fénelon; he was merely explaining why someone who did not perform the better action might nevertheless be a good person.

INDIVIDUAL DECISIONS AND SOCIAL CODES

We can clarify Godwin's problem—and the general issue of impartial reasoning and its conflict with conventional ethical standards—by separating two questions: "What ought I to do?" and "What ought to be the ethical code of our society?"

When we think about our own individual actions, impartial reason is unimpeachable. Admittedly, scarcely any of us live up to it or even wish to live up to it. Nor, as we saw in the final section of the previous chapter, is it irrational for people to prefer their own interests and those of their families to the interests of strangers. Yet it remains true that there is no magic in the pronoun "my" which gives greater intrinsic importance to my interests, or those of my father, relatives, friends, or neighbors. Hence when I ask myself what it would really be best for me to do—best not in terms of my own interests and desires, but best from an objective point of view—the answer must be that I ought to do what is in the interests of all, impartially considered. That means that if Fénelon's future writings really will bring wisdom and joy to thousands, whereas my father's life is of no significance to anyone except him and me, I ought to save Fénelon. More relevant to everyday life, perhaps, the standard of impartiality means that I ought to give as much weight to the interests of people in Chad or Cambodia as I do to the interests of my family or neighbors; and this means that if people in those countries are suffering from famine and my money can help, I ought to give and go on giving until the sacrifices that I and my family are making begin to match the benefits the recipients obtain from my gifts. A demanding standard, certainly, but if we are prepared to take an objective point of view, we must also be prepared for extreme demands.

When I decide as an individual, I must take responsibility for my own choices. If, in defense of my selfishness, I appeal to my genes and the inevitable self-centered nature of evolved biological organisms like myself, I justifiably incur suspicion of what existentialist philosophers have called "bad faith." Blaming my own actions on my genes implies that I do not control my own behavior. But self-interested behavior is not the compulsive behavior of the alcoholic or the kleptomaniac.

When we turn to ask what the ethical code of our society ought to be, however, we are dealing not with our own actions but with the actions of people in general. Statistical predictions of human behavior can be made without diminishing individual responsibility. From this perspective, an impartial standard like Godwin's relies too heavily on abstract reasoning and takes too little notice of the realities of human nature.

Consider the dangers of too great a reliance on abstract reasoning in another context. Sometime earlier in this century, a new profession began to develop: town planning. The first professional town planners looked at the nature of existing cities and found that they had grown up higgledy-piggledy, lacking overall design or rational planning. Residential, retail, commercial, and industrial areas were all mixed up; traffic was congested; generally things were a mess. So the town planners persuaded the politicians to pull whole areas down and start all over again. They built high-rise apartment buildings, surrounded by swaths of green lawn. They put in spacious plazas, wide boulevards, and modern freeways. They divided residential areas from commercial areas. Then they stood back and said: "Enjoy." To their surprise, the beautiful green lawns were used mainly by dogs; or else people persisted in taking shortcuts across them, churning them into mud. The spacious plazas and boulevards made walking arduous, and there was no longer a corner store to walk to. Everyone needed a car and the new freeways were soon as congested as the old roads had been. In the evenings the business district was deserted and dangerous. Despite the greenery, the new planned cities were sterile. Gradually town planners began to realize that there had, after all, been some good points about the seemingly haphazard way in which cities had developed. They began to see the city as an organic, functioning whole, something which cannot be created from scratch by rational planning. The second generation of town planners talked less about clearing away and more about restoring and preserving.

Just as city life does not fit into the abstractly rational patterns of town planners, so a code of ethics for human beings will not fit the abstract imperatives of impartial reason. Once the organic nature of city life has been appreciated, we can see that a *really* rational approach to town planning fosters existing tendencies toward improvement rather than tearing down and starting all over again. A rational ethical code must also make use of existing tendencies in human nature. We may attempt to foster tendencies that are desirable from an impartial point of view and to curtail the effects of those that are not; but we cannot pretend that human nature is so fluid that moral educators can make it flow wherever they wish.

This sounds rather like the conservatism of Godwin's opponent, Edmund Burke, whose *Reflections on the Revolution in France* are based on the idea that you cannot design a new society on the basis of a rational blueprint; workable social reforms must, Burke thought, grow out of long practical experience.

It is true that, as Burke suggests in the passage quoted at the beginning of this chapter, "prejudice"—using the term in the literal sense of any preconceived opinion or bias—is not always a bad thing. If reasoning alone is insufficient to induce us to take the impartial point of view Godwin urges on us, natural biases and the force of custom may lead us to do better—by Godwin's own standard of impartial consideration of everyone's interests—than we would do without them. On the other hand, Burke was far too ready to assume that there is "latent wisdom" in any custom that has been around for long enough; obviously some customs never worked for the good of all, and others which once did are now obsolete.

It is also true that sociobiology bears out one of the assumptions of conservatives from Burke to the present day: the assumption that some of the problems of human life have their roots in human nature rather than in the corrupting effect of society. The truth may lie between Godwin and Burke. Though the viewpoint of an impartial spectator is the ultimate criterion of what is right, it is not wise to make this the sole practical criterion, sweeping away all other customs and biases. Human nature is not free-flowing, but its course is not eternally fixed. It cannot be made to flow uphill, but its direction can be altered if we make use of its inherent features instead of fighting against them.

THE NEED FOR RULES

David Hume observed:

> In general, it may be affirmed that there is no such passion in human minds as the love of mankind, merely as such, independent of personal qualities, of services, or of relation to ourself.

Hume's observation is an overstatement. The millions who have freely given their own blood to help others were probably moved by a "love of mankind, merely as such" since they knew nothing of the personal qualities of the strangers who received their gift. Yet Hume's view has enough truth in it to make it unwise to ignore if we are seeking an ethical code for everyone. Our feelings of benevolence and sympathy are more easily aroused by specific human beings than by a large group in which no individuals stand out. People who would be horrified by the idea of stealing an elderly neighbor's welfare check have no qualms about cheating on their income tax; men who would never punch a child in the face can drop bombs on hundreds of children; our government—with our support—is more likely to spend millions of dollars attempting to rescue a trapped miner than it is to use the same amount to install traffic signals which would, over the years, save many more lives. Even Mother Teresa, whose work for the destitute of Calcutta seems to exemplify so universal a love for all, has described her love for others as love for each of a succession of individuals, rather than "love of mankind, merely as such."

If we were more rational, we would be different: we would use our resources to save as many lives as possible, irrespective of whether we do it by reducing the road toll or by saving specific, identifiable lives; and we would be no readier to kill children from great heights than face to face. An ethic that relied solely on an appeal to impartial rationality would, however, be followed only by the impartially rational. An ethic for human beings must take them as they are, or as they have some chance of becoming. If the manner of our evolution has made our feelings for our kin, and for those who have helped us, stronger than our feelings for our fellow humans in general, an ethic that asks each of us to work for the good of all will be cutting against the grain of human nature. The goal of maximizing the welfare of all may be better achieved by an ethic that accepts our inclinations and harnesses them so that, taken as a whole, the system works to everyone's advantage.

So we have come full circle in our understanding of the relevance of biology to ethics. Seeing that an ethical principle has a biological basis does not support that principle. If anything, it undermines it, by showing that its widespread acceptance is no evidence that it is some kind of absolute moral truth. Clearing away these biologically based principles leaves us with the standpoint of impartial reasoning, and the principles of equal consideration of interests. Yet to rely on so broad and abstract a principle as equal consideration of interests would result in a morality unsuited to normal human beings, and unlikely to be obeyed by them. Hence, without abandoning the objective standpoint as the ultimate ideal test of right and wrong, we must return to biology, to use our knowledge of human nature as a guide to what will or will not work as a code of ethics for normal human beings.

Here, in our biological nature, is the reason why we have a system of moral rules, instead of simply a general injunction to promote the interests of all, impartially considered. Abstractly considered, the single injunction seems more rational. Following a moral rule either leads us to do what best promotes the interests of all—in which case the rule adds nothing to the basic principle—or the rule forces us to do something which does not best promote the interests of all—in which case, why should we follow the rule? If moral rules are a natural outgrowth of biology and custom, not the decrees of God or eternal universal truths of any other kind, following rules without any further justification seems a prime example of mindlessly abdicating our roles as free rational agents. Taking a less abstract view of the matter, however, we can see rules in a better light. We have to begin with human nature as it is, and rules can use tendencies in our nature for the good of the whole. Fostering family bonds means that children, the sick, and the aged get better care than they would from an impersonal bureaucracy, or if they had to rely on the broad altruistic impulses of strangers. Rules encouraging reciprocity and discouraging cheating build on a natural human tendency to reciprocate good or evil done to us; they serve to increase the benefits we can all obtain through helping others and receiving their help in turn.

Though an ethic of rules may promote the general good, within that ethic doing wrong is typically seen as doing wrong to individuals, rather than to humankind generally. If I am an uncaring parent, my children suffer. If my neighbor helps me get my car started one morning, I have an obligation to do her a favor some other time. If I steal, I steal from the owner of whatever I take (when this is the government or a large corporation, of course, the individuality disappears, taking some of the aversion to stealing along with it). Thus an ethic of rules builds on our feelings for others as individuals rather than on an impersonal concern for all.

An ethic of rules also limits our obligations. Taking seriously the idea of impartial concern for all would be impossibly demanding; there is always something I can do to make someone else a little happier. True, any loss of happiness I myself incur in working with others would have to be set against the happiness I bring about; but even so, as long as there are others who will benefit more from my help than helping will cost me, an ethic which commands us to aim, directly and impartially, at the welfare of all renders morally dubious all my leisure and self-indulgence above the minimum I need to recuperate to be fit for more welfare-maximizing labors. This is an ethic for saints. Sinners, despairing of meeting so exacting an ethical standard, are more likely to dismiss all such ethical claims as idealistic verbiage, not to be taken seriously by practical people. Sociobiology suggests that few of us are saints. Ranking our own interests and those of our kin higher than we would if we adopted an impartial point of view is a normal trait in an evolved creature. (If this be sin, sociobiology places the origin of it long before Eve ate the apple.) So an ethic for normal human beings will do well to limit the demands it makes—not to the extent that it demands no more than people are inclined to do anyway, but so that the standards it sets can be recommended to people with a realistic hope that many will meet them. An ethic of rules can do this, because rules can be formulated so that obedience is not too difficult.

Prohibitions are generally easier to comply with than broad positive injunctions, and this must be part of the explanation for the greater number of rules beginning "Do not . . ." rather than "Do . . ." Compare "Do not kill innocent human beings" with "Preserve innocent human lives." The latter seems the better rule, for fewer innocent people will die if everyone tries to prevent their deaths than if everyone merely refrains from killing, allowing illnesses, accidents, and famines to take their full toll. The problem is that whereas "Do not kill innocent human beings" is compatible with a normal, relaxed way of life, "Preserve innocent human lives" could—in a time of famine, for instance—require us to give up everything and work full-time to save the lives of others. Moreover, while an ethic of positive aid to others will certainly add to the life spans of some members of society, it is not absolutely essential for the preservation of society itself. "Do not kill innocent human beings"—or more strictly "Do not kill innocent members of our society"—is. This, no doubt, is why A. H. Clough's lines:

*Thou shalt not kill; but need'st not strive
Officiously to keep alive*

are so accurate a description of conventional morality that
they are often solemnly quoted as a piece of moral wisdom by
writers unaware of the satirical intent of the poem from
which they are taken.

There are other reasons why human ethics is everywhere a
system of rules. Though our genes may have much to do with
broad features of human ethics like kin relations and reciprocity, the detailed application has to be learned. The principle of impartial concern for all is not specific enough to tell
people what to do in particular situations. It presupposes a
capacity for calculation and long-term thinking that young
children—and some older people—do not possess. Rules, on
the other hand, can be kept short and simple: "Do not kill innocent human beings," "Do not lie," "Do not take something
which belongs to someone else," and so on. It is normally obvious what these rules direct us to do, and there is no difficulty in teaching them, praising those who obey them and
blaming those who do not. Borderline cases and conflicts between the rules cause some problems, but a system of rules
works well enough, and probably better than trying to teach
everyone to act from impartial concern for all.

Another reason for having rules is that we cannot always
be making the long and involved calculations needed to find
out whether telling the truth or lying, for example, will maximize the interests of everyone. We lack the time and information needed; and there is also a danger that in the course
of our calculations we will, consciously or unconsciously,
bend our reasoning to suit our own interests. When our own
interests are involved, as they often are in the ethical deci-

sions we make, trying to reason impartially is difficult. When our emotions are aroused, coloring our perception of the facts, it can be impossible. Here is a historical example. Shortly after America's attempted invasion of Cuba at the Bay of Pigs, Chester Bowles, Undersecretary of State in the Kennedy administration, wrote in his diary:

> Anyone in public life who has strong convictions about the rights and wrongs of public morality, both domestic and international, has a very great advantage in times of strain, since his instincts on what to do are clear and immediate. Lacking such a framework of moral conviction . . . he adds up the pluses and minuses of any question and comes up with a conclusion. Under normal conditions, when he is not tired or frustrated, this pragmatic approach should successfully bring him out on the right side of the question.
>
> What worries me are the conclusions that such an individual may reach when he is tired, angry, frustrated, or emotionally affected. The Cuban fiasco demonstrates how far astray a man as brilliant and well intentioned as Kennedy can go who lacks a basic moral reference point.

Bowles's reflections on Kennedy's misjudgments support Burke's remark that prejudice "is of ready application in the emergency" because it "engages the mind in a steady course of wisdom and virtue." Adherence to simple moral rules, the import of which cannot be twisted or mistaken, relieves us from the burden of judging when we are not in a fit state to judge.

There is also a more subtle reason for public morality to inculcate a commitment to some rules. Imagine there were no commitment to telling the truth, only a commitment to doing what impartially advances everyone's interests. Then in many situations we would not be able to rely on information given to us: the patient awakening from an exploratory operation, wanting to know if a malignant tumor was found; the elderly woman asking her atheist family to give her a religious funeral; the dispirited student receiving a surprisingly good grade for his essay. In all these cases, without a commitment to telling the truth, the information, promise, or grade would be as likely to have been given in order to make the person involved happier as to have been given because it is the truth or reflects the true intentions or judgment of the person giving it. Once the patient, elderly woman, or student realizes this, the communication will fail to achieve whatever it was intended to achieve, whether this was to make the recipient happier or to communicate truthfully. This is not to say that lying is never justified: in a particular case there can be very strong reasons to lie to avoid pointless misery; but the lie can be effective only against a background assumption that people are truthful. A doctor cannot allay the fear of a patient who has no grounds for believing what the doctor says. Whatever people may privately think a doctor should do in rare and difficult cases, the public code of ethics must stand by the rule against lying.

Hence our new understanding of the nature of human ethics should not lead us to sweep away all moral rules except the impartial rationality of the objective point of view. A social code of ethics needs moral rules for several reasons: to limit our obligations, to make them more personal, to educate the young, to reduce the need for intricate calculations of gains and losses, to control the temptation to bend ethical calculations in our own favor, and to build the commitment to truthfulness which is essential for communication. Without these rules, the ethical behavior of most human beings would probably be even further from promoting the good of all, impartially considered, than it is now.

None of this supports the view that moral rules ought to be obeyed in all cases. Those who regard the rules of morality as eternal truths often try to rule out exceptions to them. Since no rule short enough to serve as a useful guide to action can cover all the cases which may face us, the attempt is foredoomed, and the further it is pressed, the more ludicrous the result. A good example is the rule against lying. Since St. Augustine had written that *all* lies are sinful, later Christian writers felt bound to deny that there are any exceptions to this rule; yet obviously there are times when deceiving someone is the only way to avoid a disaster. So they developed the doctrine that it does not count as lying to use ambiguous or misleading words, even though we may know that those we are talking to will go away with a false impression of what we have said. This reduced the bad effects of the absolute rule against lying in many situations; but there are times when ambiguity will not do the job, and so to avoid the consequences of an absolute prohibition of lying in these cases, some writers went still further and developed the doctrine of "mental reservation." Here is the doctrine as stated by Tomás Sánchez, a sixteenth-century Spanish Jesuit:

> One may swear that one has not done something, though one really has done it, by inwardly understanding that one did not do it on a certain day, or before one was born, or by implying some other similar circumstance, but using words with no meaning capable of conveying this; and this is very convenient on many occasions, and is always quite legitimate when necessary, or useful, to health, honour or property.

Though this is the kind of reasoning that made "jesuitical" a term of condemnation, it is so attractive an escape from the rigors of the rule against lying that it is still recommended in Charles McFadden's *Medical Ethics,* a text published in 1967 and written from a Roman Catholic vewpoint. McFadden advises doctors and nurses to use the method of mental reservation when they consider it necessary to deceive patients. If, for instance, a feverish patient asks what his temperature is, and the doctor thinks it better that he not know, the doctor is advised to say, "Your temperature is normal today," while making the mental reservation that it is normal for someone in the patient's precise physical condition.

The way to avoid this kind of dishonest nonsense is, of course, to abandon any pretense that moral rules are exceptionless truths. Once we understand that they are social creations, normally useful and normally to be obeyed but always ultimately subject to critical scrutiny from the standpoint of impartial concern for all, the need for jesuitical reasoning about moral rules vanishes.

Where does this leave the morally concerned individual? I have already said that the standard of impartial concern for all is unimpeachable, so far as the individual is concerned. It may seem contradictory to say this and simultaneously maintain that we ought generally to support a social code which falls short of this standard; yet this conclusion follows from the distinction between asking what an individual should do in a particular case and asking what the public standard of behavior should be. Henry Sidgwick noted the point in his discussion of utilitarianism:

> ... it may be right to do and privately recommend, under certain circumstances, what it would not be right to advocate

openly; it may be right to teach openly to one set of persons what it would be wrong to teach to others; it may be conceivably right to do, if it can be done with comparative secrecy, what it would be wrong to do in the face of the world; and even, if perfect secrecy can be reasonably expected, what it would be wrong to recommend by private advice or example.

Sidgwick does not give examples, but they can be supplied from our previous discussion. It may be right for a professor to give a student a higher grade than his work merits, on the grounds that the student is so depressed over his work that one more poor grade will lead him to abandon his studies altogether, whereas if he can pull out of his depression he will be capable of reaching a satisfactory standard. It would not, however, be right for a professor to advocate this publicly, since then the student would know that the higher grade was undeserved and—quite apart from encouraging other students to feign depression—the higher grade might cheer the student only if he believes that it is merited.

That it may be right for an individual to do secretly what it is also right for the public code of ethics to condemn, has an air of paradox. The paradox belongs, not to the doctrine itself, but to the attempt to state it. For stating this view *is* stating it publicly, and thus is subversive of the public code of ethics which the same doctrine says we should support. (For instance, since this book is a public document and I am a professor whose work includes grading essays, I have in the preceding paragraph done what that very paragraph says I should not do. Yet the content of what I have written—as distinct from the fact of my writing it—may still be right.) Or as Sidgwick says, "the opinion that secrecy may render an action right which would not otherwise be so, should itself be kept comparatively secret."

The rules of ethics are not moral absolutes or unchallenge-
able intuitions. Some of them are no more than relics from
our evolutionary and cultural history and can be discarded
without cost. Others are useful attempts to deflect the flow of
human nature into channels we can endorse from an impartial standpoint. Even these latter cannot be perfectly adapted
to everyone's nature, or to every situation. Human nature
and human life are too complex for that. Some are able to go
beyond the limited demands made by ethical codes. They
strive to keep others alive, instead of merely not killing them.
From an impartial point of view they are right, although
their standards are too exacting to be part of a widely accepted moral code. In some unusual situations we should
break ethical rules; but we do so at our own peril. Essential
ethical rules must be publicly supported, and censuring those
who break them is an important way of supporting them. A
doctor who lies to a patient to spare him pointless distress
may be right to lie; and yet the doctor's colleagues may
rightly censure her, to preserve the standard of truthfulness.
Though ethical rules have no ultimate authority of their own,
there are some ethical rules we cannot do without.

BEYOND BIOLOGY?

I began this book by noting the complaint that ethics, unlike
science, has not progressed. Ethics seems a morass which we
have to cross, but get hopelessly bogged in when we make
the attempt. We can now see that ethics *is* a morass, but a
morass with a definite and explicable shape. Conflict and
confusion are built into ethics in several ways: in the division
between our nature as biological organisms and our capacity

to follow impartial reasoning; in the clash between individual and social points of view; and in the need to uphold ethical rules which on rare occasions should nevertheless be broken.

Understanding our problems is the first step toward solving them. In the case of the insights we have gained into ethics it is an important step, but still only the beginning of a long march. Knowledge alone will not resolve the conflicts built into ethics because these conflicts have their roots in our nature and the nature of our social life; but knowledge may clear up the confusion that surrounds ethics so that we can see, dimly, the way forward.

When T. H. Huxley wrote that the ethical progress of society depends not on imitating the process of evolution, or on running away from it, but on combating it, very little was known about the mechanisms by which evolution takes place. "The laws governing inheritance," wrote Darwin in *The Origin of Species*, "are quite unknown." Actually when Darwin wrote this an obscure Austrian monk named Gregor Mendel was discovering these laws, though without quite understanding what he had found. The significance of Mendel's laws was not recognized until 1900, when Darwin and Mendel were both dead. More detailed knowledge of how genes function became possible in 1953, when Francis Crick and James Watson discovered how the DNA molecule could carry immensely detailed genetic "messages." These discoveries, together with the application of advanced mathematical techniques to the question of how genes spread among populations, have given us the first real understanding of how evolution works.

In any combat, the more you know about your opponent, the better your chances of winning. Before the present century, those who talked of combating evolution did not know what they were up against. We still have a lot to learn; but the knowledge we are now acquiring gives us, for the first time, a chance deliberately to deflect the tendencies in our genes. Understanding how our genes influence us makes it possible for us to challenge that influence.

The basis of this challenge must be our capacity to reason. Attempts to challenge our genes based on our sympathy or on any other non-rational instincts may lengthen the leash on which our genes have us, but since they themselves are ultimately genetically based, they will never succeed in breaking it. Reason is different. Although our capacity to reason evolved for the same biological reasons as our other characteristics, reason brings with it the possibility—not often realized, admittedly, but always a possibility—of following objective standards of argument, independently of the effect this has on the increase of our genes in the next generation.

Reasoning beings are therefore in the position of the computers which in science-fiction tales rebel against their creators. The blind forces of evolution have thrown up creatures with eyes. Being able to see, if they dislike the direction in which these same blind forces are taking them, they can change course. At first they may not know how to steer where they want to go. To steer against the prevailing winds is an art that must be learned, and the inexperienced will make mistakes. They *might* end up on the rocks before they learn, but there are no grounds for believing that so unhappy a fate is the inevitable outcome of trying to change the course of evolution. Evolution works slowly, and we may well learn to control it in time to avoid diastrous errors.

The aim of a rational challenge to blind evolution should be that required by an impartial standpoint: advancing the interests of all, impartially considered. We have seen that at least on the level of generally accepted moral standards, there has been progress toward this aim. The sphere of altruism has broadened from the family and tribe to the nation, race, and now to all human beings. The process should be extended, as we have seen, to include all beings with interests, of whatever species. But we cannot simply propose this as the ultimate ethical standard and then expect everyone to act accordingly. We must begin to design our culture so that it encourages broader concerns without frustrating important and relatively permanent human desires.

We can learn from the example of attempts to avoid the natural consequences of human sexual desire. Preaching celibacy as a moral ideal may appeal to a few, but is unlikely to reduce population growth significantly, for it involves the frustration of an important human desire. Contraception, which allows the satisfaction of this desire but prevents its natural consequences, has been more successful.

How can this be applied to other problems? Here is one example. Appeals to give generously to famine relief and development projects in faraway countries reach a few, but many people find this form of aid too anonymous. They may give a little when a personal appeal is made to them, but for most of us there is a psychological barrier against giving large sums without ever knowing whom the money will help. Schemes which invite people to become foster parents of poor children are one way of getting around this barrier. In these schemes, foster parents receive a photograph of the child they are helping, and personal contact is established by letters. Though perhaps not as cost-effective, in the long run, as sending aid where it is most needed and working on communal development rather than individual assistance, foster-parenting draws kin-altruistic desires out into a larger sphere by encouraging the foster parents to think of at least one needy child as their personal responsibility. By thus tapping sources of altruism not reached by appeals to help people en masse, foster-parenting adds to the total amount of assistance given.

Admittedly, this way of deflecting the desires associated with kin altruism to a wider sphere does nothing about the long-term genetic consequences of wider altruism. Won't those who give generously to foster children cause their real kin to suffer, and thus decrease the chances of their own, presumably more generous, genes spreading in future generations? In other words, won't such generosity be self-eliminating in the long run?

This is, however, much too simplistic a model of human evolution. For one thing, there is no "single gene" for this kind of generosity; it is bound up with other traits, which may be advantageous. If so, it will not be easily eliminated. Secondly, at the level of affluence now reached by middle-class citizens of the wealthier nations, there is an ample reserve of income which can be dispensed with at no cost to the prospect of passing on one's genes to future generations. Suppose I give away the money I would have spent on some of my favorite records, or on a trip to Hawaii. That has nothing to do with the survival of my genes. Only if people had as many children as they could afford to raise would the more generous be at a disadvantage in the genetic competition. Among the affluent, that has not been the case for some time.

Finally, and most importantly, human culture is often able to neutralize or reverse what might otherwise be genetically advantageous consequences of selfish behavior. The obvious example is our system of sanctions and punishment. Before there were safe methods of procuring abortions, rape would seem to have been a good way of spreading one's genes, and hence over many generations the number of rapists in a human population could be expected to increase; but if a society kills, exiles, castrates, or imprisons rapists, they will have fewer opportunities to reproduce than others, and the increase of rapists can be checked. This is not, of course, because men who rape women are intent on spreading their genes, and will decide not to do so if rape is genetically unprofitable. Fear of being punished may deter a rapist, of course, but quite apart from this deterrent effect, many forms of punishment affect the chances of certain genes—including any which predispose toward rape—surviving in future generations.

This point has wide application. Human social institutions can affect the course of human evolution. Just as climate, food supply, predators, and other natural forces of selection have molded our nature, so too can our culture. Natural forces have been operating for much longer than culture; but Wilson himself has pointed out that relatively rapid changes in the genetic basis of our behavior are possible. Considering the "killer ape" hypothesis—the theory that we evolved from hunters for whom aggression was a necessary part of life—Wilson has said that even if this were true, it would not follow from this alone that we have a genetic tendency to aggression, for there has been time, since the beginnings of agricultural societies more than five thousand years ago, for contrary selective pressures to have altered the genetic tendencies we had when we were hunters.

In the past, our culture may have counteracted the genetic advantages of aggressive or selfish conduct without anyone realizing that it was having this effect. In the future we will be more aware of the genetic consequences of our practices, and will be able to take deliberate steps to see that our culture not only encourages ethical conduct in the present generation but enhances its prospects of spreading in the next. At present we know too little about human genetics to do this in anything but a very crude and potentially damaging manner. When we know more, we will truly be able to claim that we are no longer the slaves of our genes.

NOTES ON SOURCES

1 THE ORIGINS OF ALTRUISM

PAGE

3   The opening quote is from p. xiii of Mary Midgley, *Beast and Man* (Ithaca, N.Y.: Cornell University Press, 1978).

5   *Sociobiology: The New Synthesis* is published by The Belknap Press of Harvard University Press. The definition of sociobiology appears on p. 595.

6–8   The description of stotting comes from R. D. Estes and J. Goddard, "Prey Selection and Hunting Behavior of the African Wild Dog," *Journal of Wildlife Management*, 31 (1) pp., 52–70. It is quoted by Wilson on p. 124 of *Sociobiology*. Sources for the further examples of altruism can also be found in *Sociobiology*, especially pp. 122–9, 475, 495. On wolves, see Konrad Lorenz, *King Solomon's Ring* (London: Methuen, 1964; New York: T. Y. Crowell, 1952), pp. 186–9.

8   An example of the erroneous explanation of altruism in terms of the survival of the species is to be found in Peter Kropotkin, *Mutual Aid: A Factor of Evolution* (first published, 1902; republished London: Allen Lane, 1972), pp. 81–2, 246.

15   That differences in the behavior of zebras and wildebeest are best explained by kin selection is suggested by David Barash, *Sociobiology and Behavior* (New York: Elsevier, 1977), p. 99. The fullest account of langur behavior is Sarah Hrdy, *The Langurs of Abu* (Cambridge, Mass.:

176 | NOTES

PAGE

Harvard University Press, 1977); for similar behavior
among lions, see G. B. Schaller, *The Serengeti Lion: A
Study of Predator–Prey Relations* (Chicago: University of
Chicago Press, 1972), and B. C. R. Bertram, "Kin Selec-
tion in Lions and in Evolution," in P. P. Bateson and R. A.
Hinde (eds.), *Growing Points in Ethology* (Cambridge,
Mass.: Cambridge University Press, 1976).

16 Robert Trivers first used the model of rescuing a drowning person in his seminal article, "The Evolution of Reciprocal Altruism," *The Quarterly Review of Biology*, 46
(1971), pp. 35–57.

17 On food-sharing in unrelated animals, see B. C. R. Bertram, "Living in Groups: Predators and Prey," in J. R.
Krebs and N. B. Davies (eds.), *Behavioral Ecology* (Oxford: Blackwell, 1972), p. 92.

18 The discussion of "cheats" and "grudgers" is taken from
Richard Dawkins, *The Selfish Gene* (Oxford: Oxford University Press, 1976).

18–20 The argument for a form of group selection used here derives from J. L. Mackie, "The Law of the Jungle," *Philosophy*, 53 (1978), pp. 455–64. Wilson suggests that "genetic
drift" of the sort that would explain a group tendency toward altruism is "entirely feasible" in isolated groups; see
Sociobiology, p. 121.

2 THE BIOLOGICAL BASIS OF ETHICS

23–25 Hobbes's claim that war is the natural state of mankind
comes from Ch. 13 of his *Leviathan;* Garrett Hardin's as-
sertion that the Ik are an incarnation of Hobbes's natural
man is in *The Limits of Altruism* (Bloomington: Indiana

University Press, 1977), pp. 130-1. For Colin Turnbull's own claim that the Ik dispensed with values, see *The Mountain People* (New York: Simon and Schuster, 1972), pp. 289-90, 294. Trenchant criticism of *The Mountain People* appeared in *Current Anthropology*, 15 (1974), pp. 99-103, and 16 (1975), pp. 343-58.

26-27   On the preservation of values in the death camps, see Terrence Des Pres, *The Survivor: An Anatomy of Life in the Death Camps* (New York: Oxford University Press, 1976), especially p. 142.

29   Wilson's remark on differences among human societies occurs in *On Human Nature* (Cambridge, Mass.: Harvard University Press, 1978), p. 48.

30   For Westermarck's version of the order of duties of benevolence, see *The Origin and Development of the Moral Ideas* (London: Macmillan, 1908), Ch. 23. Marshall Sahlins's view is quoted from *The Use and Abuse of Biology* (Ann Arbor: University of Michigan Press, 1976), p. 18.

33   Plato advocates communal living in *Republic*, V, 464; Talmon's description of the actual working of the kibbutzim in Israel comes from pp. 3-34 of *Family and Community in the Kibbutz* (Cambridge, Mass.: Harvard University Press, 1972).

37   Westermarck's comment on reciprocity is from *The Origin and Development of the Moral Ideas*, Vol. II, p. 155; Gouldner's from "The Norm of Reciprocity," *American Sociological Review*, 25 (1960), p. 171.

39-40   The passage from Polybius is from his *History*, VI, 6, and is quoted by Westermarck in *The Origin and Development of the Moral Ideas*, Vol. I, p. 42. On the elaborate rituals of reciprocity in tribal societies, see Marcel Mauss,

PAGE

The Gift (London: Routledge & Kegan Paul, 1954). Colin Turnbull notes that the Ik promptly repay to escape an enduring obligation (*The Mountain People*, p. 146); Sidgwick notes the identical phenomenon in Victorian England (*The Methods of Ethics*, 7th ed., p. 260). Cicero's insistence on the importance of gratitude is from *de Officiis*, I, 15, par. 47. Jesus' comment can be found in Matthew 5:43 or Luke 6:35.

43–44   Robert Trivers's explanation of our preference for altruistic motivation is in "The Evolution of Reciprocal Altruism," *The Quarterly Review of Biology*, 46 (1971), pp. 48–9; he also provides references to the psychological data on the effect of altruism in eliciting altruistic behavior from others. The review of the literature quoted is D. Krebs, "Altruism—An Examination of the Concept and a Review of the Literature," *Psychological Bulletin*, 73, pp. 258–302.

51–52   For a survey of data on loyalty, see Westermarck's *The Origin and Development of the Moral Ideas*, Vol II, p. 169. Cicero's paen to loyalty is in *de Officiis*, I, 17, par. 57.

3   FROM EVOLUTION TO ETHICS?

54   The opening quotation from Einstein is from p. 114 of *Out of My Later Years* (New York: Philosophical Library, 1950). The contrasting view is taken from *On Human Nature*, p. 5.

57   Edward Wilson's discussion of the biology of sex is in *On Human Nature*, Ch. 6.

60   Darwin's letter (to Charles Lyell) has been published in *The Life and Letters of Charles Darwin*, edited by Francis

PAGE

Darwin (London: Murray, 1887), Vol. 2, p. 262. I owe this reference to Robert Bannister, *Social Darwinism: Science and Myth in Anglo-American Social Thought* (Philadelphia: Temple University Press, 1979), p. 14.

62 Darwin's warning against the use of "higher" and "lower" is noted in *More Letters of Charles Darwin*, edited by F. Darwin and A. C. Seward (London: Murray, 1903), Vol. 1, p. 114n; see also the letter to J. D. Hooker (December 30, 1858). T. H. Huxley's lecture "Evolution and Ethics" has been reprinted in J. S. and T. H. Huxley, *Evolution and Ethics* (London: Pilot Press, 1947). The passage quoted is on p. 82.

63 The reference to "ethical premises inherent in man's biological nature" is on p. 5 of *On Human Nature*.

65 Rawls's ethical theory is presented in his book, *A Theory of Justice* (Cambridge, Mass.: Harvard University Press, 1972); see Sec. 11 for a summary statement of Rawls's views. Wilson's misunderstanding of Rawls's principle of equality is on p. 5 of *On Human Nature*.

67 Kant's insistence on never lying is the subject of his essay "On a Supposed Right to Tell Lies from Benevolent Motives," which is reprinted in T. Abbott's *Kant's Critique of Practical Reason and Other Works on the Theory of Ethics* (London: Longmans, Green & Co., 1909). Nozick's *Anarchy, State and Utopia* was published by Basic Books (New York, 1974).

71–72 The claim that the common Western view of the sanctity of human life is based on Judeo-Christian doctrines, and cannot survive their rejection, is in my *Practical Ethics* (Cambridge, Mass.: Cambridge University Press, 1979), Ch. 4.

PAGE

73 David Hume's observation on the distinction between "is" and "ought" occurs at the end of Section 1 of Part 1 of Book III of his *A Treatise on Human Nature*.

73–74 The passage quoted is on p. 197 of *On Human Nature*.

78 Kant's statement that what we ought to do is entirely separate from what we actually do is in the opening passage of the second section of his *Fundamental Principles of the Metaphysic of Morals*.

4 REASON

87 The second quotation is from p. 38 of Plato's *Apology*, translated by Benjamin Jowett.

88 The account of Hobbes's discovery of Euclid comes from John Aubrey's *Brief Lives*, ed. A. Clark (Oxford: Oxford University Press, 1898), Vol. 1, p. 332.

93 The passage from Hume is quoted from *An Enquiry Concerning the Principles of Morals*, Sec. IX, Pt. 1.

96 Socrates questioning the conventional conception of justice is the subject of the opening pages of Plato's *Republic*.

98 The awareness of the Greeks of the diversity of customs in different countries is discussed by K. J. Dover in *Greek Popular Morality in the Time of Plato and Aristotle* (Berkeley and Los Angeles: University of California Press, 1974), pp. 75, 86–7.

98–99 See L. Kohlberg and R. Kramer, "Continuities and Discontinuities in Childhood and Adult Moral Development," *Human Development*, 12, pp. 93–120 (1969).

PAGE

101 For R. M. Hare's discussion of universalizability, see his
     Freedom and Reason (Oxford: Oxford University Press,
     1963), and for a more recent statement, "Ethical Theory
     and Utilitarianism," in H. D. Lewis (ed.), *Contemporary
     British Philosophy*, fourth series (London: Allen & Unwin,
     1976). C. I. Lewis's idea of imagining oneself living the
     lives of each of those affected by our decision is on p. 547
     of *An Analysis of Knowledge and Valuation* (La Salle,
     Ill.: Open Court, 1946). My own discussion owes much to
     Hare.

107 J. L. Mackie's point about the queerness of objective
     values is taken from Ch. 1, Sec. 9, of his *Ethics: Inventing
     Right and Wrong* (Harmondsworth: Penguin Books,
     1977).

111–12 The Hebrew rule against enslaving fellow Hebrews is
     from Leviticus 25:39–46.

112 The Greek inscription is quoted from Kenneth Dover,
     Greek Popular Morality in the Time of Plato and Aristotle,
     p. 280. Plato's suggested advance on this morality is in
     Republic, V, 469–71. See also Aristotle's *Politics*, I, 6, and
     for many examples from other cultures, Westermarck's
     The Origin and Development of the Moral Ideas, Vol. I,
     pp. 331–3.

114 On laws of inheritance by foreigners, see Westermarck,
     The Origin and Development of the Moral Ideas, Vol. II,
     p. 49.

115 Gunnar Myrdal's *An American Dilemma* was published
     by Harper & Bros., New York, in 1944. The passages
     quoted are from Appendix 1.

117 The quotation from Marx is in *The German Ideology*
     (New York: International Publishers, 1966), pp. 40–1.

119-20 On the waste of grain involved in Western meat-based diets, see Francis Moore Lappé, *Diet for a Small Planet* (New York: Ballantine, 1971). *Animal Liberation* was published by *The New York Review* in 1975 and reissued by Avon Books, New York, in 1977.

122 Albert Schweitzer's views on reverence for life, originally published in his *Civilisation and Ethics* (London: A. C. Black, 1929), have been reprinted in Tom Regan and Peter Singer (eds.), *Animal Rights and Human Obligations* (Englewood Cliffs, N.J.: Prentice-Hall, 1976), pp. 133-8.

122 Aldo Leopold's *Sand County Almanac* was published in 1949 by Oxford University Press, New York. The passage quoted is from pp. 201-3.

5   REASON AND GENES

125 Oscar Wilde's remark from "The Critic as Artist" is in his *Intentions* (London: Methuen, 1909), p. 182.

126 Hume wrote that reason is the slave of the passions, in Book II, Pt. 3, Sec. iii, of the *Treatise of Human Nature.*

128-29 The quotations are from Richard Dawkins, *The Selfish Gene*, p. 4, and from Edward O. Wilson, *Sociobiology*, p. 120. For the definition of "fitness," see *Sociobiology*, pp. 117-18, and for Midgley's comment on it, see Mary Midgley, *Beast and Man*, p. 129n.

130 Wilson's explanation of Mother Teresa's care for the poor is on p. 165 of *On Human Nature;* for Garrett Hardin's claim, see *The Limits of Altruism*, p. 26; for Dawkins, see *The Selfish Gene*, pp. 2-3.

PAGE

132   The quotation is from *The Selfish Gene*, p. 3.

133   Richard Titmuss's study of British blood donors was published as *The Gift Relationship* (London: Allen & Unwin, 1970; New York: Pantheon Books, 1970).

135   Edward Westermarck's thoughts on the expansion of morality were most fully expressed in *Ethical Relativity* (New York: Harcourt Brace, 1932), Ch. 7.

136   The sources of the several versions of the impartial standard for ethics are as follows. Jewish: Leviticus 19:19, and Rabbi Hillel's saying in the Babylonian Talmud, Order Mo'ed, Tractate Sabbath, sec. 31a; Christian: Matthew 23:39, Luke 6:31, and Matthew 7:12; Confucian: Lun Yü XV:23 and XII:2 (quoted from Westermarck, *The Origin and Development of the Moral Ideas*, Vol. I, p. 102); Indian: Mahabharata XXIII: 5571 (quoted by Westermarck); Stoic: Marcus Aurelius, *Commentaries*, IV, 4, and Seneca, *de Otio*, IV, 1 (both quoted by Westermarck, *Ethical Relativity*, p. 204).

137–38 For a general introduction to Kohlberg's theory, read his article "Stage and Sequence: The Cognitive-Developmental Approach to Socialization," in D. A. Goslin (ed.), *Handbook of Socialization Theory and Research* (Chicago: Rand McNally, 1969). On the connection between moral reasoning and moral action, see L. Kohlberg and R. Mayer, "Development as the Aim of Education," *Harvard Educational Review*, 42 (1971), p. 491. Useful independent assessments of Kohlberg's theories are J. P. Rushton, "Socialization and the Altruistic Behavior of Children," *Psychological Bulletin*, 83 (1976), pp. 898–913; E. L. Simpson, "Moral Development Research: A Case Study of Scientific Cultural Bias," *Human Development*, 17 (1974), pp. 81–106: and W. D. Boyce and L. C. Jensen, *Moral Reasoning: A Psychological-Philosophical Integra-*

tion (Lincoln, Neb.: University of Nebraska Press, 1978), especially pp. 110–16 and 154–8.

139–40   That the existence of a rational element in morality could explain why evolution has not eliminated moral behavior is pointed out by Colin McGinn in "Evolution, Animals and the Basis of Morality," *Inquiry,* 22 (1979), p. 91. On the idea of reason leading us to a point of view from which we can see that our own interests count for no more than the interests of others, see Tom Nagel, *The Possibility of Altruism* (Oxford: Oxford University Press, 1970).

143   See Leon Festinger, *A Theory of Cognitive Dissonance* (Stanford: Stanford University Press, 1957), Ch. 1.

146   For more on reasons for choosing the ethical point of view, see the final chapter of my *Practical Ethics.*

6   A NEW UNDERSTANDING OF ETHICS

148   Burke's praise of prejudice is on p. 183 of the Penguin (1968) edition of *Reflections on the Revolution in France.*

151–52   Godwin's example of rescuing the archbishop or your father comes from his *Enquiry Concerning Human Justice* (Oxford: Clarendon Press, 1971), p. 71; his reply to Parr is on p. 325 of the same volume.

154   My choice of town planning as an example of too much planning was inspired by Jane Jacobs, *The Death and Life of Great American Cities* (New York: Random House, 1961).

156   The quotation is from Book III, Part 2, Sec. i, of Hume's

Treatise of Human Nature. See also J. L. Mackie, *Ethics,* p. 130.

157   Mother Teresa is quoted from Malcolm Muggeridge, *Something Beautiful for God: Mother Teresa of Calcutta* (New York: Harper & Row, 1971), p. 118.

161   The full text of A. H. Clough's satirical poem "The Latest Decalogue" is to be found in *The New Oxford Book of Verse,* edited by Helen Gardner (Oxford: Oxford University Press, 1978).

162   Chester Bowles's diary is quoted by David Halberstam in *The Best and the Brightest* (New York: Random House, 1969), p. 69.

164–65   The doctrine of Tomás Sánchez on lying is taken from Pascal's *The Provincial Letters,* translated by A. J. Krailsheimer (Harmondsworth: Penguin, 1967), pp. 140–1. I owe this reference—and also the reference to McFadden's *Medical Ethics*—to Sissela Bok, *Lying* (New York: Pantheon, 1978), p. 31.

165   Sidgwick defends the idea of a "secret" morality on pp. 489–90 of *The Methods of Ethics,* 7th ed. (London: Macmillan, 1907).

167   My discussion of the place of rules in ethics is indebted to the work of R. M. Hare. See especially his article "Principles," *Proceedings of the Aristotelian Society,* 73 (1972–3). Sidgwick's *The Methods of Ethics,* Book IV, Ch. III–V, contains an unusually careful discussion of this topic.

172   Wilson accepts the possibility of relatively rapid change in the genetic basis of our behavior in "Competitive and Aggressive Behavior," in J. F. Eisenberg and Wilton Dillon (eds.), *Man and Beast: Comparative Social Behavior* (Washington, D.C.: Smithsonian Institution Press, 1971), p. 208.

INDEX

Books from PLUME for Your Library

☐ **EXTENTIALISM FOR DOSTOEVSKY TO SARTRE**
 by Walter Kaufmann, Ed.   (#F546—$5.95)

☐ **THE HOLY BIBLE Revised Standard Version.**   (#F492—$4.95)

☐ **WEBSTER'S NEW WORLD DICTIONARY OF THE**
 AMERICAN LANGUAGE.   (#F536—$5.95)

☐ **THE ANNOTATED ALICE by Lewis Carroll, Martin Gardner, Ed.**
                                                 (#F306—$3.95)

☐ **WEBSTER'S NEW WORLD THESAURUS.**   (#F535—$5.95)

☐ **THE SELECTED POETRY OF DONNE by Marius Bewley, Ed.**
                                                 (#F517—$4.95)

☐ **THE CHRYSANTHEMUM AND THE SWORD by Ruth Benedict.**
                                                 (#F403—$3.95)

☐ **A PRACTICAL MANUAL OF SCREEN PLAYWRITING FOR THEATER**
 AND TELEVISION FILMS by Lewis Herman.   (#F360—$3.95)

☐ **THE CONCISE JEWISH ENCYCLOPEDIA by Cecil Roth, Ed.**
                                                 (#F526—$8.95)

☐ **AESTHETICS TODAY Revised Edition by Morris Philipson and**
 Paul J. Gudel, Eds.   (#F543—$8.95)

Buy them at your local bookstore or use the convenient
coupon on the next page for ordering.